Medicinal and Environmental Chemistry: Experimental Advances and Simulations (Part I)

Edited by

Tahmeena Khan
Integral University
Department of Chemistry
India

Abdul Rahman Khan
Integral University
Department of Chemistry
India

Saman Raza
Isabella Thoburn College
Department of Chemistry
India

Iqbal Azad
Integral University
Department of Chemistry
India

&

Alfred J. Lawrence
Isabella Thoburn College
Department of Chemistry
India

Medicinal and Environmental Chemistry:
Experimental Advances and Simulations *(Part I)*

Editors: Tahmeena Khan, Abdul Rahman Khan, Saman Raza, Iqbal Azad and Alfred J. Lawrence

ISBN (Online): 978-981-4998-27-7

ISBN (Print): 978-981-4998-28-4

ISBN (Paperback): 978-981-4998-29-1

need for a court order if at any point you breach any terms of this License Agreement. In no event will any delay or failure by Bentham Science Publishers in enforcing your compliance with this License Agreement constitute a waiver of any of its rights.

3. You acknowledge that you have read this License Agreement, and agree to be bound by its terms and conditions. To the extent that any other terms and conditions presented on any website of Bentham Science Publishers conflict with, or are inconsistent with, the terms and conditions set out in this License Agreement, you acknowledge that the terms and conditions set out in this License Agreement shall prevail.

Bentham Science Publishers Pte. Ltd.
80 Robinson Road #02-00
Singapore 068898
Singapore
Email: subscriptions@benthamscience.net

BENTHAM SCIENCE

CONTENTS

FOREWORD

Environmental pollution (air, water and soil) and human health are inextricably linked. The developing countries are engaged in a wide range of activities that are causing enormous damage to the environment, ecosystems that sustain both our species and Earth's legacy of biodiversity, and human health. If our society takes constructive actions now, or at least soon, it will not be too late to prevent or repair many of these important environmental problems, which threaten the welfare of people and most other species. A more respectful attitude toward the natural world is also urgently needed, for the world is one family, "Vasudhaiva Kutumbakam".

This innovative book will attract scientists interested in environmental pollution and human health with a view to offer remediation techniques. The book chapters have been authored by experts from their fields, both scientists and academicians, and would benefit the readers.

Viney P. Aneja
North Carolina State University
USA

PREFACE

With the drastic disturbance in environmental harmony and balance, there has been a rise in global deaths and diseases, calling for the exploration of novel remediation strategies for innovative drug action mechanisms and target identification. The fine balance between human and ecological health is getting disturbed, leading to serious implications, including the occurrence of new pathogens and diseases, the novel coronavirus SARS-CoV-2, being the most recent instance having gripped the entire globe.

Environmental diseases are non-communicable and are caused by chronic exposure to toxic pollutants. Other contributory causes of environmental diseases include radiation, pathogens, allergens and psychological stress. Their increasing occurrence is due to industrialization, changes in farming protocols and the increase in exposure to chemicals released into the environment. Lifestyle changes, including the increased use of tobacco and processed foods, also greatly contribute to the environmental/lifestyle diseases burden.

Though separately medicinal chemistry and environmental chemistry have been widely explored, yet their close association and interdependence have been overlooked. By exploring the association between these two focal areas, the present book aims to provide solutions and curative strategies for the well-being of humans and the environment as a whole.

The ten chapters included in this book are focused on diverse topics trying to blend the fields of environmental chemistry and medicinal chemistry and have been authored by experts, scientists and academicians from renowned institutions. A wide range of topics has been explored in the book to make it relevant to environmental chemists and students. The chapters have been designed so as to introduce environmental contaminants and techniques for their quantification and removal. Also, a medicinal perspective for remediation of environmental hazards, from therapeutic strategies available to the design of new and safer drugs, is introduced through experimental and simulation approaches.

Specialized chapters have been dedicated to persistent organic pollutants, heavy metals, and plastics, which have become a major source of pollution, along with their remediation. The effect of environmental xenoesterogens on human health has been discussed in one chapter, while in another, the potential of natural curing agents to combat ecotoxicity has been explored. To further elaborate the importance of safe chemical practice, the concept of green chemistry has also been introduced.

As we are aware that drug discovery for a particular disease is a time taking endeavour, therefore a few chapters have also been dedicated to in-silico predictions like molecular docking and virtual models for biological properties, the software used and their utility in making futuristic and accurate predictions to make drug discovery efficient, quicker and cost-effective. Chapters summarizing the challenges of medicinal chemistry as well as the advances of biomolecular simulations for drug designing with respect to ecotoxicity are also included.

The book will prove beneficial for academicians, students of environmental chemistry, pharmacy, researchers, scientists, computational chemists, pharmacologists, environmentalists, policymakers and postgraduate students. It would also provide researchers and medicinal chemists the information regarding the latest research done and the modern techniques used to develop more effective and safer drugs that would not be harmful to the environment. In this way, the proposed book would be highly beneficial to the audience it hopes to cater to.

Tahmeena Khan
Integral University
Department of Chemistry
India

Abdul Rahman Khan
Integral University
Department of Chemistry
India

Saman Raza
Isabella Thoburn College
Department of Chemistry
India

Iqbal Azad
Integral University
Department of Chemistry
India

&

Alfred J. Lawrence
Isabella Thoburn College
Department of Chemistry
India

List of Contributors

Ahmad I.	Isabella Thoburn College, Lucknow, India
Ahmad M.	Zakir Husain College of Engineering and Technology, Aligarh Muslim University, Aligarh, India
Alam Z.	Shibli National PG College, Azamgarh, India
Ali A.	Zakir Husain College of Engineering and Technology, Aligarh Muslim University, Aligarh, India
Ansari J.A.	King George's Medical University, Lucknow, India Shibli National PG College, Azamgarh, India
Azad I.	Integral University, Lucknow, India
Bajpai S.	Amity University, Lucknow, India
Bhatia S.	Isabella Thoburn College, Lucknow, India
Bhatia S.	Isabella Thoburn College, Lucknow, India
Biswas K.	Indian Institute of Technology Kanpur, Kanpur, India
Gupta A.	CSIR-Central Institute of Medicinal and Aromatic Plants (CSIR-CIMAP), Lucknow, India
Gupta N.	CSIR-Indian Institute of Toxicology Research, Lucknow, India
Jabeen F.	Jazan University, Jazan, Saudi Arabia
Khan A. R.	Integral University, Lucknow, India
Khan M.A.	K.K.L.K.M, Kathara, Kanpur, India
Khan T.	Integral University, Lucknow, India
Khare A.	Indian Institute of Technology Kanpur, Kanpur, India
Kumar S.	CSIR-Indian Institute of Toxicology Research (CSIR-IITR), Lucknow, India
Mahdi A. A.	King George's Medical University, Lucknow, India
Mishra A.	Indian Institute of Information Technology, Prayagraj, India
Mishra N.	Indian Institute of Information Technology, Prayagraj, India
Mulpuru V.	Indian Institute of Information Technology, Prayagraj, India
Nagar P.K.	Indian Institute of Technology Kanpur, Kanpur, India
Nasibullah M.	Integral University, Lucknow, India
Patel D.K.	CSIR-Indian Institute of Toxicology Research (CSIR-IITR), Lucknow, India
Rahman Q.I.	Integral University, Lucknow, India
Raza S.	Isabella Thoburn College, Lucknow, India
Sharma M.	Indian Institute of Technology Kanpur, Kanpur, India
Sharma P.	Babasaheb Bhim Rao Ambedkar University, Lucknow, India
Sharma V.P.	CSIR-Indian Institute of Toxicology Research (CSIR-IITR), Lucknow, India
Singh N.	Amity University, Lucknow, India

Verma J.   CSIR-Indian Institute of Toxicology Research (CSIR-IITR), Lucknow, India

Yadav A.   Indian Institute of Technology Kanpur, Kanpur, India

Environmental Chemistry: Applications, Interactions and Paradigm Shift in Futuristic Approaches

Vinod Praveen Sharma[1,*], **P. Sharma**[2] and **Abdul Rahman Khan**[3]

[1] *CSIR-Indian Institute of Toxicology Research (CSIR-IITR), Lucknow, India*
[2] *Babasaheb Bhim Rao Ambedkar University, Lucknow, India*
[3] *Integral University, Lucknow, India*

Abstract: Environmental chemistry is an interdisciplinary science with multiple importance in the dynamic lifestyle and consumption pattern. Globally, the environmental regulatory agencies and research institutions feel the extreme need for environmental chemistry for the identification of the nature, source, monitoring, and remediation of pollutants. The pollutants may range from heavy metals, organometallics, polycyclic aromatic hydrocarbons, and nutrients, to the runoff of various other contaminants, their transportation, and interaction with living organisms. Their rapid and accurate separation, identification, quantification using sophisticated techniques, characterization, and understanding of the interactions and mechanisms are the key components of analytical chemistry, for better biochemical or physiological understanding. Contaminants generally have short or long-term toxic implications on the surrounding environment due to direct impact or through bioactivity. Management of environmental pollutants, with minimal impact on biodiversity and human population, is the desired objective of most of the Research & Development programs of International societal relevance. The coordination and effective implementation through sustainable, green, computational technologies may provide the best strategic solutions to the innovators, academicians, and stakeholders, amidst constraints on resources.

Keywords: Characterization, Environmental, Management, Strategic, Sustainable, Technologies.

INTRODUCTION

We need to attempt visualising the pathways and draw a roadmap for a sustainable future. We all know that newer materials/ alloys will lead to improved products, and novel processes may improve manufacturing efficiency and reduce

---

* **Corresponding author Vinod Praveen Sharma:** CSIR- Indian Institute of Toxicology Research, Mahatma Gandhi Marg, Lucknow, Uttar Pradesh; E-mail: vpsitrc1@rediffmail.com

Tahmeena Khan, Abdul Rahman Khan, Saman Raza, Iqbal Azad and Alfred J. Lawrence (Eds.)

energy usage, waste generation, and resultant pollution. The success of civilisation may thus depend on the ability to create such newer materials and novel applications [1 - 4]. It is projected that in emerging economies the production and sales of chemicals will continue to grow rapidly irrespective of the pandemic-related challenges. It will be going through a period of mergers, acquisitions, and several types of restructuring. Chemicals have a major role to play in global resource flow and value chain. China (approx. 37% of global sales) and the European Union (16% of global sales) remain the highest users of chemical products, followed by the United States and BRICS countries (Brazil, Russia, India, China, and South Africa). R&D experts feel that the pharmaceutical industry is highly innovative and competitive, with dependency on research funds, and is subject to strong government regulations. Moreover, the pharmaceutical industry has been shifting towards developing primary care and small-molecule medicines, transitioning to specialty medications for ageing populations.

In the post-Covid-19/ SARS-CoV-2 situations, it is expected that the thickly populated urban areas will always be susceptible to diseases, which may spread *via* airborne pathogens, surfaces, and human-to-human contact. They will be faced with immediate and long-term challenges [8 - 15]. The basic information collected and compiled regarding the dynamic properties of macromolecules may propel a shift to structural bioinformatics, from understanding single structures to analysing conformational ensembles. The molecular dynamic simulations have now evolved into a mature technique, which helps to understand the structure-activity relationship of macromolecules. Moreover, it helps in providing better insights into biological actions like enzyme mechanisms, regulation, transport across membranes, and building of large structures, such as ribosomes, viral capsids, transcriptions, *etc.*

INDUSTRIALIZATION AND CHEMICAL RESEARCH WORKS

Industrialization and globalisation are undergoing a paradigm shift and with an abundance of raw materials and comparatively economically priced manpower, our country is privileged to take the benefit of cost-effective manufacturing. Thomas Kuhn motivated for a change in interdisciplinary approaches originating from natural sciences and applied chemistry, with the utilization of computational techniques. Chemical research must aim to support new radical approaches and ground-breaking projects, through investigators who are exceptional leaders in terms of the originality and significance of R & D contributions.

Agency for Toxic Substances and Disease Registry (ATSDR) is an organisation for the dissemination of best solutions based on R & D findings, for trustworthy information with direct relation to health, to prevent harmful exposures and preventing diseases associated with toxic substances [1]. It also offers an emergency response program for societal benefits, at a global level. Human exposures may be associated with chemicals or mixtures which are toxic and may originate from environmental and occupational sources. The exposures may be from other vital sources and drugs or indoor air pollutants and affect susceptible populations, communities, or indirectly associated tribal inhabitants.

PARADIGM CHANGES AND INNOVATIONS

In connection with the paradigm changes towards sustainability, the intensification of global agriculture practices must be interconnected with objectives that are directed to meet customer demands for resilience and biosphere protection. We need to take steps for eradicating hunger and at the same time securing food for an increasing global population of nine to ten billion, by 2050. This may require steady growth in food production amidst potential global environmental risks. The regulation of the usage of agrochemicals and synthetic fertilizers requires appropriate coordination and cooperation with agriculturists, institutions involved in R&D, civil society, and governments. All of this will become more critical in a global scenario of resource constraints, health hazards, and the dynamic requirements of a growing population. Modern technology, and its competent use for mitigating future pandemics or environmental crises, is the most effective tool we have in our arsenal to protect communities.

We must take a lesson from the reduction in industrial activities and emissions during the recent lockdown; the government's restrictions on movement from one area to another has led to a significant reduction in global pollution levels, and rejuvenation of nature. This was evident from various publications of reputed journals of science and technology. It has affected societies and economies around the globe and is expected to reshape the activities of professional life. The crisis fallout is both amplifying familiar risks and creating new avenues for managing systemic challenges to build an improved climate for coming generations in a universal scenario. Efforts are needed for a reduction in the release of irritating toxic gases affecting the pulmonary system. It is vital to address niche areas holistically *viz.* drug development, trade, governance, health, education, and labour, to mention the few where the balance of risk and opportunities exists after strength, weaknesses, opportunities, and threats (SWOT) analysis. We may utilise artificial intelligence and machine learning to provide momentum to a series of economically feasible activities in the field of environmental chemistry, with intelligent analysis applications. Preparedness, thus, becomes the strongest weapon in the anticipated disasters or natural calamities. Revolutionary analytical chemistry and computational biology, with better insight, will have a great future. The passive sampling devices, *in situ* methods, and specially designed assay techniques may serve as an improved tool for environmental chemistry.

Chemists and scientists need to produce critically needed medical supplies using 3D printing to cope with the increasing pressure for personal protective equipment(s), *viz.* N95 masks, face shields, hands-free door openers, other healthcare products, food supplies, antiviral Active Pharmaceutical Ingredient (APIs) synthesis *etc.* Significant challenges exist for testing appliances/accessories, vaccines, scalability of production, and novel packaging solutions. Another aspect that is concerning environmental experts is the persistence of coronavirus in the environment. It is anticipated that growth in chemical-intensive industries may create potential risks, depending on technology selection and usage of chemicals. The futuristic era may also create opportunities for innovation towards improved production processes and safer nanomaterial or biobased packaging products for drugs and other materials in maximum demand.

The knowledge of environmental chemistry helps to predict the behaviour of matter and its efficacy, in addition to monitoring or synthesizing diverse types of materials. Several activities have been directed toward supporting industrial processes and creating new products and materials, *viz.* formulations of new pharmaceutical products, creation of polymers, designing of fertilizers, neem-based pesticides to increase food production, transforming bitumen into automobile fuel, *etc.* We need to understand the implications of the transformation of materials, and other chemistry practices, for the sustainability of existing ecosystems. Most of the chemicals released into the environment during the manufacture and use of many products enter into our bodies through the air, water, food, and skin; nowadays there are > 80,000 chemicals with known or suspected adverse health effects [2, 4, 7].

Green chemistry is based on a set of principles aimed at the reduction or elimination of hazardous substances from the design, manufacture, and application of chemical products. It moves products and processes toward an innovative economy based on renewable feedstocks and toxicity is intentionally prevented at the molecular level.

AGRICULTURAL CHEMISTRY AND GREENER HABITATIONS: INTERLINKS

Broadly speaking, sustainable agriculture seeks to achieve three goals: farm profitability, community, and environmental stewardship. We need to manage the agricultural farms and watersheds using appropriate strategies and practices to maintain biophysical stability, critical feedstock, and carbon sinks in soils and biomass. The information digital technologies, artificial intelligence, and big data applications are making unprecedented strides in our daily lives and integrating with societal changes of products and services with disproportionate connectivity dependent phenomenon. The smart green cities of the future may be surrounded and intertwined with ecological infrastructure systems, which may be constantly monitored through sensors, robotic systems, and multifaceted drones or digital technologies. Artificial intelligence may refer to computer systems that may sense climate changes, weather conditions, and specific environments, and act with dynamic responses.

Bio-derived adsorbents may serve as a realistic technology for the benign recovery of diffuse elements from liquid effluents and hydrometallurgy processes (Table **1**). It is being explored as a strategy beyond the remediation of heavy metals and pollutants, by utilising biosorption within a circular economy, for the cycling of precious and critical metals in higher-value applications.

Table 1. Salient Bio-derived Adsorbents.

S.No.	Bio-derived Adsorbents	Relevant Details	Remarks
1.	Hydrogel	A Potassium polyacrylate-based super absorbent polymer	Used for reducing soil erosion and irrigation frequency, limiting nutrient loss, increasing the yield of crops. The hydrogel polymer may swell upon contact with water molecules, absorbing up to 400-800 times its weight and then acting as a controlled water release system by releasing water only upon the plant's requirement.
2.	Activated Charcoal	Granular activated charcoal is produced from coconut shell charcoal by the process of steam activation.	Coconut shell-based granule activated carbon, silver-impregnated charcoal for water treatment filtration media under gravel aquarium filter-chemical filtration for saltwater and freshwater.
3.	Absorbent Polymer	Water retaining super absorbent polymer for soil mixing and hydroponics gardening manure	Serves as a super absorbent, polymeric product for soil and hydroponics.
4.	Cellulose-based adsorbents	Cellulose-based adsorbents from sunflower, grass waste, jackfruit, cucumber, orange, and sweet lime (Mosammi) peels, spent tea leaves, *etc.*	-
5.	Cellulose fibrils	The green preparation of cellulose fibrils from palm leaf-stalk fibres for hydrogel applications	The green preparation of cellulose fibrils from palm leaf-stalk fibres for hydrogel applications

(Table 1) cont.....

S.No.	Bio-derived Adsorbents	Relevant Details	Remarks
6.	Hydrophilic polymers- a combination of psyllium seed mucilage and Stockosorb® *i.e* STS for Sweet basil	Application of super absorbent polymer and plant mucilage improved essential oil quantity and quality of Ocimum basilicum.	Ocimum basilicum (sweet basil) is a valuable medicinal plant that is sensitive to water deficit, and water shortage negatively affects sweet basil yield and quality. Water availability in the root zone of basil could ameliorate the negative effects of water shortage. Investigators explored the effects on water use efficiency when using Stockosorb® (STS) and psyllium seed mucilage (PSM) as hydrophilic polymers (HPs) and the effects of these HPs on essential oil quality, quantity, and yield.

FUTURISTIC APPROACHES

The future of plastics and polymers significantly depends on the synergy with nano-technology and revolution in composites, for their sustainability, multi-functionality, and applications. However, the main challenges faced by plastics industrial units are the non-degradability of waste and pollution in oceans and beaches and the impact on flora and fauna. In recycling units, there are issues of lack of infrastructure, stringent legal bindings, and shortage of trained manpower with competency as per requirements. The environmental load of the polymer-wastes is a grave global challenge. With these challenges to tackle in the coming twenty years or so, the implementation strategy, specifically for the polymers and composites industry, may include technology development, value chain improvements, retention of talented workforce, availability of appropriate funds, and policy support.

In spite of several efforts, we still need to study the gaps and conduct holistic detailed research for understanding the nature and significance of environmental exposures of several environmental pollutants. The pollutants of concern include new chemical entities, agrochemicals, antibiotics, pharmaceuticals, nutraceuticals, surfactants, flame-retardants, polycyclic hydrocarbons, *etc.*

BIOMATERIALS

Biomaterials need to be safe and biocompatible, with increased performance efficiency. With the advancement of time, the interface between multi-disciplinary technologies may change and the ambit may surpass the domain of devices, embracing regenerative medicines, cell therapeutics, tissue engineering, gene delivery, personal healthcare products, *etc.* With a changing scenario of population, prosperity, prospecting, and environmental protection, it is anticipated that the discovery, processing, and usage of newer materials and finished innovative products may change in the upcoming decades, based on consumer expectations and demands. The carbon dioxide levels may reach twice the pre-industrial level by 2050. With higher concentration, it may lead to greater global warming and rising sea levels due to the melting of glaciers and the release of methane in the Tundra. We need to conserve water to avoid drought in the coming future. The explosion of the human population and the concomitant need and greed has made us exploit nature which has led to an imbalance in the relationship between humans and the environment. We need to change our behaviour and lifestyle for the benefit of the next generation.

ENVIRONMENTAL CONSCIOUSNESS AND CARBON FOOTPRINT REDUCTION

Our best ecological footprint will be to reduce pollution and make the best endeavour for land and water utilisation in an efficient manner, composting the green waste materials, recycling, and converting the waste to energy. We must be environmentally conscious and focus on sustainability approaches. The renewable energy resources, *viz.* wind turbines, solar panels, and biogas, should be used to the optimum levels. We must guard the integrity of nature for continued growth, adopt humane behavioural approaches with clean energy sources, and sufficiently reduce fossil fuel consumption. We need a universe wherein nature and inhabitants thrive to conserve the environment and also fulfill and enrich the lives of others. We have the responsibility to preserve and respect the delicate balance of nature with nurturing of the planet for the future, to assure prosperity for all. Our responsibility is also to communicate with stakeholders and address environmental challenges.

NEW CHEMICAL ENTITIES AND STRUCTURE-ACTIVITY RELATIONSHIPS

The quantitative structure-activity relationship studies are being used in environmental sciences for complementing the experimental data obtained from *in silico, in in vitro*, or *in vivo* studies, associated with environmental results and extrapolation. This accelerates the interpretation and utilisation of data for societal usages. It provides better knowledge of the effect of contaminants and determines the fate of toxicants. We may attempt to integrate the state of art databases and predictive models for the development of tools to assess the risks, implications, and safety assurance of new chemical entities (NCEs) or moieties. The sustainable development goals (SDGs) are well-defined and accepted goals and targets, designed for guiding the envisaged policies and practices of several contributing countries. The International Union for Pure and Applied Chemistry (IUPAC) has identified the importance of safety activities for personnel training and capacity building with a radical change in attitudes to encourage or help the partners by improving both safety and security. Accidents or chemical disasters may be prevented with fundamental knowledge of first aid and better handling practices [4 - 15].

It is a challenge to the fast-expanding population. The anaerobic microbially mediated technologies may also serve as an economical alternative to physical and chemical processes for both sanitation and resource recovery purposes. The computer-assisted methods help in estimating the properties of substances, evaluating the fate determination processes, and contribute to predictive toxicology, using computational methods and bioinformatics. Quantum mechanics are also finding great relevance in understanding the chemical structure and fate of molecules. Most of the tools and updated methods are acceptable by international organizations of repute, such as ATSDR, EPA, ISO, OECD, and WHO, for cross-referral studies (Table **2**).

Table 2. Important Guidelines for Quality Implementation in Manufacturing and Testing.

S. No.	Guidelines/ Specifications and Website	Relevant Details
1	Agency for Toxic Substances and Drug Registry https://www.atsdr.cdc.gov/	ATSDR
2	Good Manufacturing Practices https://en.wikipedia.org/wiki/Good_manufacturing_practice	GMP
3	Good Laboratory Practices https://dst.gov.in/ngcma	GLP
4	Current Good Manufacturing Practices (CGMPs) for Food and Dietary Supplements https://www.fda.gov/food/guidance-regulation-food-and-dietary- supplements/current-good-manufacturing-practices-cgmps-food-and-dietary-supplements	CGMP
5	ISO/IEC-General requirements for the competence of testing and calibration laboratories https://www.iso.org/home.html	ISO 17025:2017
6	The Organisation for Economic Co-operation and Development Guidelines https://www.oecd.org/chemicalsafety/testing/good-laboratory-practiceglp.htm	OECD
7	World Health Organisation https:www.who.int	WHO

INTERFACE BETWEEN MEDICINAL AND ENVIRONMENTAL CHEMISTRY

The health consequences may be based on interactions among biological systems and the environmental disturbance, amidst increasing human populations. There are close interrelationships between medicinal chemistry and environmental chemistry in the area of climate change and unexpected diseases due to unknown viruses and unregulated biological phenomena. These disciplines are the intersection of biochemistry and have transboundary implications, including a circular economy and social strings. Several important molecules are designed to serve as bioactive molecules. Drug development is a tedious process involving characterization, standardization, validation, clinical trials, toxicity evaluations, bioefficacy, compatibility, regulatory approvals, *etc.* with high attrition rates, sufficient expenditure, and long timeframes. Varied stoichiometric ratios of reagents may result in different varieties of products with concentration variances based on structure-activity relationships. The complete process of drug development or manufacturing may be subdivided into a series of operational processes, *viz.* milling, granulation, coating, tablet preparations, packaging, labelling, transportation, and marketing. Scientists are using the knowledge of bioactivities, computational chemistry, chemical biology, enzymology, shelf life, *etc.* of natural products for the development of new therapeutic agents. The quality aspects of pharmacy-based formulations and medicines are aimed to assure the fitness of medicinal products in concurrence with Pharmacopoeias and European Union or United States Food and Drug Administration guidelines. Regulatory agencies like Environmental Protection Agency (EPA), through the Toxic Substances Control Act (TSCA), helps to produce safe chemicals, safeguard health, and regulate harmful substance usage. Recently on June 5th, 2020, EPA has proposed the use of Inpyrfluxam, which is a pyrazole carboxamide fungicide, for foliar and seed treatment in the agriculture sector [4 - 20].

Intelligent packaging is an integral part of norms and includes the associated factors, *viz.* assurance of the efficacy of the drug, patient safety, intended shelf-life, uniformity in drug even in varied production lots, quality control checks implementation, and documentation of materials and processes involved, as per Good Manufacturing Procedures requirements of OECD. Packaging and safe labelling focus on dispensing, dosing, sterility, display of technical information, precautions, *etc.* as per regulations. Several medicinal products may be sensitive and affected by environmental conditions. Thus, it is vital to store and transport them as per the directions of the manufacturers. They need to be distributed with the utmost care and if directed, the cold chain process is strictly implemented to maintain quality.

The product safety management or implementation of the quality management system, as per IS/ISO/IEC 17025:2017, Organisation for Economic Cooperation and Development (OECD), Good Laboratory Practices, and World Health Organisation (WHO), protocols and guidelines are vital for validation, compliances and safety assurance [1 - 10, 13, 18 - 20].

We need novel therapeutic agents, sensors, and technologies for the treatment of upcoming diseases and complications. Moreover, a better understanding of factors and pathways for the prevention of complications in environmental health and safety risks is important. The aspect of antibiotics resistance and the overburden of toxicants is being studied by researchers at a global level. The issue of the presence of medicinal complexes, pharmaceutical active ingredients or agents in water bodies or ecosystem is complex. The investigations of preclinical studies may generate sufficient data related to pharmacokinetics and toxicity. The innovative researchers are attempting to target specificity and pathways to understand how an innovative drug may have efficacy in the treatment of diseases. The drug developers generally focus on therapeutics with a well-understood mechanism of action, minimal or low toxicity, and risk assessment, based on dose-response relationships. The bioavailability concentrations are complex in dynamic processes, depending on the chemical structure, persistence and physical/chemical properties of active ingredients in the environment.

Nowadays, the target-fishing approaches of a small molecule are important in medicinal chemistry for identifying the most probable targets, as well as virtual screening tools. For example, Chronic Obstructive Pulmonary Disease (COPD) is characterized by progressive obstruction of airflow and is due to harmful particles or gases in the lung parenchyma. There are currently no specific treatments for this and smoking cessation remains the most effective therapeutic intervention. The characteristics properties of nanocarriers, *viz.* liposomes, polymeric nanoparticles, micelles, and bioconjugates, have been explored to enhance drug solubility, dissolution, and bioavailability. The most important advantage offered by nanotechnology is the ability to specifically target organs, tissues, and individual cells, which minimizes the associated side-effects and improves the therapeutic index of drug molecules.

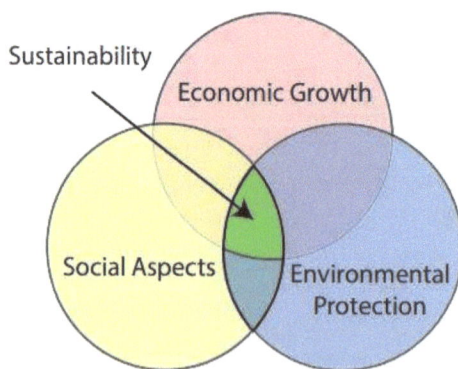

Fig. (1). Interrelationship of Sustainability to Environmental Aspects with Growth and Social Aspects.

INFERENCES

For innovation in environmental sciences, we need capacity building for the generation of a trained workforce and effective processes with a healthy work culture that leverages workforce diversity and adopts new strategies. Drug development is vibrant and has transformed globally and we need to address the challenges of cross-disciplinary issues and quality enforcement. We need concerted efforts by industrialists, academia, governments, and Non-Government Organisations (NGOs) to connect administrative, cultural, and technical experts and devise innovative solutions, including big data and machine learning. We may contribute to driving solutions in the environmental arena, technological developments, drug discovery, and translational toxicological research to ensure a sustainable future. The technologies may significantly increase the pace of research processes that may be implemented for the benefit of society (Fig. **1**).

CONCLUDING REMARKS

Several environmental improvements may be attained through non-toxic and environment-friendly chemicals. The chemical characterisation, homogeneity, stability profiling, and knowledge of master safety data sheets for individual chemicals, have profound relevance in manufacturing units or laboratories. Endocrine-disrupting chemicals are the best examples to understand the possible interventions in different matrices or use in fast-moving consumer products.

The computational models are being used to predict the possible pathways of degrading products. We need to shift our perspective towards green chemistry. It is elaborated through academic groups that instead of enacting another law that bans or regulates a chemical or a molecule which has toxicity or environmentally destructive effect, we need to enact for a product that does not have that effect.

The complexity of chemicals, environmental situations, and variability in biodiversity make direct measurements difficult for researchers. The prediction and risk assessment of the chemical properties of a chemical is important to safeguard the environment, flora, and fauna, *etc.*

CONSENT FOR PUBLICATION

Not Applicable.

CONFLICT OF INTEREST

The author confirms that this chapter contents have no conflict of interest.

ACKNOWLEDGEMENT

Declared none.

REFERENCES

[1]   Holler J. Agency for toxic registry of substances disease registry. J Environ Health 2013; 76: 46-7.
        [PMID: 24288850]

[2]   Anyamba A, Chretien JP, Small J, Tucker CJ, Linthicum KJ. Developing global climate anomalies
        suggest potential disease risks for 2006-2007. Int J Health Geogr 2006; 5: 60.
        [http://dx.doi.org/10.1186/1476-072X-5-60] [PMID: 17194307]

[3]   European Union System for the Evaluation of Substances. 2017. http://ec.europa.eu/jrc/en /scientific-
        tool/european-union-system-evaluation-substances

[4]   International Union of Pure and Applied Chemistry IUPAC. 2019.https://iupac.org/100/stories/safety-

[5]   Blotevogel J, Mayeno AN, Sale TC, Borch T. Prediction of contaminant persistence in aqueous phase:
        a quantum chemical approach. Environ Sci Technol 2011; 45(6): 2236-42.
        [http://dx.doi.org/10.1021/es1028662] [PMID: 21332222]

[6]   Salter-Blanc AJ, Bylaska EJ, Lyon MA, Ness SC, Tratnyek PG. Structure–activity relationships
        forrates of aromatic amine oxidation by manganese dioxide. Environ Sci Technol 2016; 50(10): 5094-
        102.
        [http://dx.doi.org/10.1021/acs.est.6b00924] [PMID: 27074054]

[7]   Dodson JR, Parker HL, García AM, *et al.* Bio-derived materials as a green route for precious & critical
        metal recovery and re-use. Green Chem 2015; 17: 1951-65.
        [http://dx.doi.org/10.1039/C4GC02483D]

[8]   Gallardo K, Castillo R, Macilla N, *et al.* Biosorption of rare earth elements from aqueous solutions
        using walnut shell Frontiers in Chemical Engineering 2020.

[9]   Alam MN, Islam MS, Christopher LP. Sustainable production of cellulose-based hydrogels with
        superb absorbing potential in physiological saline. ACS Omega 2019; 4(5): 9419-26.
        [http://dx.doi.org/10.1021/acsomega.9b00651] [PMID: 31460032]

[10]   Nirmala A. Thirupathaiah, Guvvali. Hydrogel/superabsorbent polymer for water and nutrient
        management in horticultural crops-review. Int J Chem Studies 2019; 7: 787-95.

[11]   OECD. 2020.https://www.oecd.org/coronavirus/en/

[12]   Saxena SK. Coronavirus Disease 2019 COVID-19, (2020), Epidemiology, Pathogenesis, Diagnosis,
        and Therapeutics, 1st ed.; Springer:Singapore, 2020.

[13]   Application of Super Absorbent Polymer and Plant Mucilage Improved Essential Oil Quantity and
        Quality of Ocimum basilicum var. Keshkeni Luvelou

[14]   Beigi S, Azizi M, Iriti M. Application of Super Absorbent Polymer and Plant Mucilage Improved
        Essential Oil Quantity and Quality of *Ocimum basilicum* var. Keshkeni Luvelou. Molecules 2020;
        25(11): 2503.
        [http://dx.doi.org/10.3390/molecules25112503] [PMID: 32481510]

[15]   Puzyn T, Leszczynski J, Cronin MTD. Recent Advances in QSAR Studies: Methods and Applications.
        1st ed., Dordrecht: Springer 2010.
        [http://dx.doi.org/10.1007/978-1-4020-9783-6]

[16]   U.S. Environmental Protection Agency. 2017. Chemistry Dashboard   https://comptox.epa.gov
        /dashboard/

[17]   United Nations Economic and Social Affairs Population Division. 2014.

[18]   Weston AD, Hood L. Systems biology, proteomics, and the future of health care: toward predictive, preventative, and personalized medicine. J Proteome Res 2004; 3(2): 179-96.
[http://dx.doi.org/10.1021/pr0499693] [PMID: 15113093]

[19]   World Health Organization. 2020.https://www.who.int/publications/i/item/clinical-management--f-covid-19

[20]   Lee Y, von Gunten U. Advances in predicting organic contaminant abatement during ozonation of municipal wastewater effluent: reaction kinetics, transformation products, and changes of biological effects. Environ Sci Water Res Technol 2016; 2: 421-42.
[http://dx.doi.org/10.1039/C6EW00025H]

CHAPTER 2

Medicinal Chemistry: Opportunities and Challenges

Jamal Akhtar Ansari[1,2,*], Abbas Ali Mahdi[1] and Zafar Alam[2]

[1] *King George's Medical University, Lucknow, India*

[2] *Shibli National PG College, Azamgarh, India*

Abstract: Medicinal chemistry is a modern branch of the pioneer subject chemistry. Medicinal chemistry is primarily associated with drug discovery and design in search of New Drug Entities (NDEs). There are different sources, such as natural and synthetic products, animals, marine invertebrates, microorganisms, and recombinant DNA approaches which have been recognized as potential reservoirs for bioactive compounds or drugs. Medicinal chemistry has made several technological innovations, such as computational chemical biology, trial-and-error approach, and bioinformatics, which have greatly improved and accelerated the efficient and competent drug development process. Although with hi-tech innovations in medicinal chemistry, there are several diseases for which treatment is still not available, including the very recent dreadful occurrence of novel coronavirus (COVID-19), which originated from Wuhan city of China. At present, there is no vaccine or drug to cure it. Moreover, the drug development process starting from the identification of a new chemical entity (NCE) to the regulatory approval of NDE is relatively complex, costly, and time-consuming. It can take 10−15 years or even longer to develop and design an NDE. The present chapter intends to discuss and emphasize the different drug sources and drug development processes in medicinal chemistry along with understanding the associated opportunities and challenges.

Keywords: Animals, Artificial intelligence, Bioinformatics, Computer-aided drug design, Challenges, Drug design, Drug development, Drug repositioning, High throughput screening, Medicinal chemistry, New chemical entity, Natural products, Opportunities, Recombinant DNA technology, Synthesis.

* **Corresponding author Jamal Akhtar Ansari:** Department of Biochemistry, King George's Medical University, Lucknow, India and Department of Chemistry, Shibli National PG College, Azamgarh, India; E-mail: jamalakhtarindia@gmail.com

Tahmeena Khan, Abdul Rahman Khan, Saman Raza, Iqbal Azad and Alfred J. Lawrence (Eds.)
All rights reserved-© 2021 Bentham Science Publishers

INTRODUCTION

The International Union for Pure and Applied Chemistry (IUPAC) in 1974 has defined Medicinal chemistry as, *"Medicinal chemistry concerns the discovery, the development, the identification and the interpretation of the mode of action of biologically active compounds at the molecular level. Emphasis is put on drugs, but the interest of the medicinal chemist is not restricted to drugs but includes bioactive compounds in general. Medicinal chemistry is also concerned with the study, identification, and synthesis of the metabolic products of drugs and related compounds"* [1]. The word drug is derived from the French word 'drogue', which means a dry herb, and it is defined as 'any substance used for diagnosis, prevention, relief or cure of some disease in a man or animal' [2]. Medicinal chemistry is primarily associated with drug discovery and design which creates a stimulating link between many scientific disciplines in search of New Drug Entities. Moreover, the field of medicinal chemistry has been revolutionized by several technological innovations, such as computational tools and bioinformatics, which have speeded up drug developments and design procedure. Drug discovery and design is the process through which potential new medicines are identified which are further used for the treatment of different diseases. It involves a wide range of scientific disciplines, including biology, chemistry, and pharmacology. Medicinal chemistry is concerned with this interaction, focusing on the organic and biochemical reactions of drug substances with their targets.

Medicinal chemistry is an interdisciplinary science combining the chemical sciences with life sciences and medical sciences at the interface to develop a potent drug. It involves biochemistry, pharmacology, molecular biology, genetics, immunology, pharmacokinetics, and toxicology on one hand, while on the other hand there are physical chemistry, crystallography, spectroscopy, computational techniques, data analysis, and data visualization [3]. The discovery of an NDE has always depended on creative and rational thinking, high-quality science and serendipity. An NDE is expected to meet an unmet medical condition or therapy where treatment modalities are not available or because of the incompatibility of existing drugs. The purpose of this book chapter is to demonstrate and discuss the possible drug sources, drug development process, advancements in drug discovery, and a robust understanding of how the medicinal chemist instigates the right experiments/strategy with the opportunities and challenges in medicinal chemistry.

SOURCES OF A DRUG

World Health Organization (WHO) describes a drug as any substance used in a pharmaceutical product that is intended to modify or explore the physiological systems or pathological states for the benefit of the recipient [4]. A drug is a substance or product that affects the physiology or pathology of living cells. In simple words, it is used as a medicine to diagnose, cure, and prevent the occurrence of a disease and disorder and prolongs the lives of patients suffering from serious or incurable diseases.

Before the twentieth century, the main sources of drugs were plants. Later, microorganisms and minerals were also recognized as potential sources of drug candidates. Nowadays, most of the drugs are obtained from synthetic, semi-synthetic and biosynthetic sources. Nature has served as a potent source of all medicaments and has been continuing since ancient times as an important source of novel bioactive compounds. These bioactive molecules are used either directly as medicinal agents or act as leads for synthetic structural modifications and optimization. There are mainly six sources, *viz.* plant, animal, mineral/earth, microbiological, semi-synthetic/ synthetic, and recombinant DNA technology (Fig. **1**) which are the potential sources of drugs which will be elaborated in detail in this chapter along with the challenges associated with these sources.

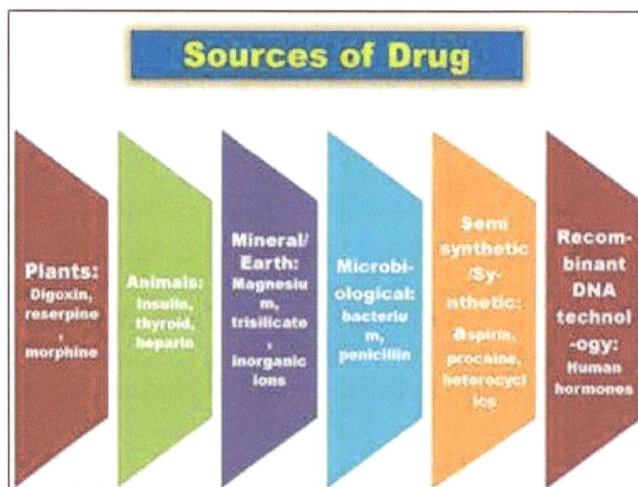

Fig. (1). Sources of a drug.

SYNTHETIC SOURCES

Synthesis is a core part of drug discovery and development. The birth of organic synthesis goes back to the early part of the 19th century, as marked by the serendipitous preparation of urea, a naturally occurring organic compound, from ammonium isocyanate, an inorganic compound. The case of medicinal chemistry and process chemistry, as it applies to drug discovery and development, is perhaps the most compelling demonstration of the impact of synthetic organic chemistry on society. Modern drug discovery and development rely on a biology-chemistr-

-medicine partnership in which biologists identify and validate a biological target relevant to the disease being addressed and synthetic chemists prepare large collections of small organic molecules to find lead compounds that bind and modulate the function of the biological target.

The organic synthesis offered the introduction of NDEs after a lot of meticulously performed experiments, satisfying the rigorous criteria which should be compatible with the desired condition *i.e.*, unmet medical need, market availability, and high attrition rate. The complexity and involvement of the multidisciplinary research process in the development of an NDE are represented in (Fig. **2**).

Organic synthetic chemists make the most creative demand for '**Lead generation**' and '**Lead optimization**'. In lead generation, several molecules are identified that have the potential to be developed into NDE. Drug discovery usually begins with target identification *i.e.*, selection of a biochemical mechanism that is involved in a disease condition. Then up to 10,000 drug candidates (NCEs) are put through a rigorous screening process to assess their interaction with the drug target. In this screening process, initial '**hits**' (NCEs active against the target) are identified. These lead compounds *i.e.*, 'hits' are then optimized through a reiterative process of pre-formulation studies, molecular drawing, synthesis, drugability, and biological and pharmacological evaluations, to find more promising clinical candidates. Lead generation is the route through which a variety of compounds that have the potential to be developed into a drug are identified and selected. Moreover, creativity is not only demanded in synthesizing desired novel bioactive molecules, but the molecules must also have suitable physicochemical properties for the means of administration, very little or no toxicity, bioavailability, must be absorbed and distributed efficiently at an adequate concentration to the desired site of action, and suitable excretion from the body *i.e.* the molecules must have ideal Absorption, Distribution, Metabolism, and Excretion (**ADME**) properties. Overall, this process of drug development takes 10-15 years to make a compatible drug for the unmet medical need. Medical practitioners undertake clinical trials and examinations to determine toxicity and efficacy to cure or improve the intended medical condition. A typical representation of the drug discovery process is depicted in Fig. **2**.

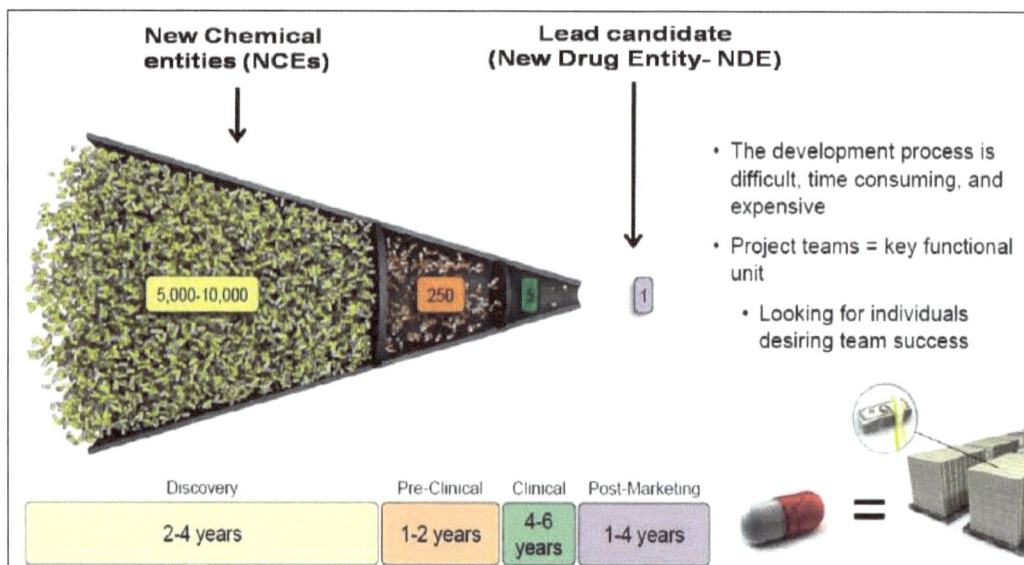

Fig. (2). Drug development process of New Drug Entity (NDE): Opportunity and Challenges.

NDEs in organic synthesis are aimed at the formation and establishment of bioactive, safe, and novel targeted molecules. Small synthetic chemical entities are preferable for oral administration.

The organic synthesis must be focused on the synthesis of biologically active compounds which show an array of biological activities. Computational research and genomic research conducted to identify the target protein, and high throughput screening to unveil the potentiality of the test molecule, have increased the chances of identifying new NDEs.

Fig. **3** represents the challenges faced during the discovery of NCEs in the research and development (R&D) process. There are various parameters such as toxicity of NCE, less clinical efficacy, clinical safety, drugability, bioavailability, pharmacokinetic properties, *etc.* causing a high degree of attrition of NDEs. Also, most of the drugs fail during the clinical studies in humans and do not satisfy the unmet medical need. This is the main reason behind the non-availability of treatment/drug/vaccine for many unmet clinical conditions/diseases, including COVID-19. Synthetic chemistry is not directly involved in the failure, but the inappropriate biological properties of NCE are responsible for the low success rate. The only way to sort out the problem is to design, synthesize and optimize new lead molecules.

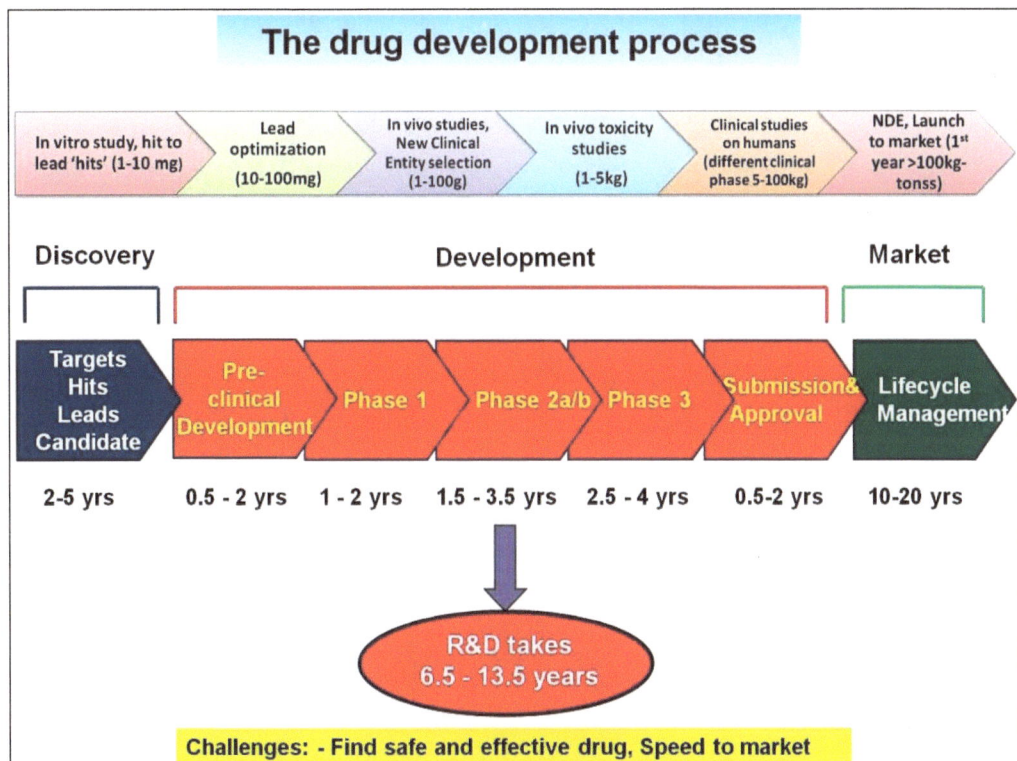

Fig. (3). The drug development process and compound requirement for R& D process: Medicinal chemistry opportunities and challenges.

Even though synthetic chemistry has been considered as the main source of a novel class of bioactive molecules, recently, analogue-based drug design is an emerging concept in medicinal chemistry. The analogy has an important role in applied sciences and is a crucial strategy in medicinal chemistry. But it is not a simple research method and involves a combination of modern *in- silico* and in-solution experimental methods. Several stand-alone drugs are considered as standard, taken as starting/reference points for this approach [5].

The protein binding site in biological targets where drug molecules interact has a complex three-dimensional structure. It is almost determinative that chiral compounds may effectively bind to these sites, and hence synthetic methods facilitating optically active pure molecules offer key synthetic armamentarium. It has been reported that 68% of drugs belonging to the top 200 brands and 62.5% of the top 200 generic drugs are in optically pure form; moreover, amongst the ten highest-selling drugs in the US in 2004, nine were chiral molecules, having a share of $53.5B in the market [6], thereby highlighting the importance of stereochemistry in drug development and design.

In this context, it is important to discuss the attempts that have been made to distinguish between drug-like and non-drug like molecules. The most familiar approach is ***Lipinski's 'Rule of Five'*** [7] that predicts a high probability of success or failure due to drug-likeness for molecules complying with 2 or more of the following parameters:

- Molecular weight is less than 500 Dalton.
- High lipophilicity (expressed as LogP less than 5).
- Less than 5 hydrogen bond donors.
- Less than 10 hydrogen bond acceptors.
- Molar refractivity between 40-130.

Besides, other approaches have emphasized the importance of parameters such as the number of rotatable bonds and polar surface area (PSA) [8].

It is widely accepted that synthetic chemistry has played a vital role in drug discovery and design which has provided incredible opportunities to medicinal chemists and pharmaceutical companies. Once researchers identify a promising compound for drug development, they conduct experiments to gather information and troubleshoot challenges, such as:

- How is it absorbed, distributed, metabolized, and excreted?
- Potential benefits and mechanisms of action.
- The best dosage.
- The best way to administer the drug (such as orally or by injection).
- Side effects or adverse events can often be referred to as toxicity.
- How it affects different groups of people (such as by gender, race, or ethnicity) differently?
- How it interacts with other drugs and treatments.
- Its effectiveness as compared with similar drugs.

PLANT-DERIVED SECONDARY METABOLITES/PHYTOCHEMICALS

Natural products have historically proven their value as a source of molecules with therapeutic potential. Over the past decade, traditional herbal medicines have acquired global importance, making an impact on both world health and international trade. Ethno-medical plants have played a key role in the healthcare system of large proportions of the world's population [9]. It is precisely the chemistry of natural products, which has fostered new developments (Fig. **4**). It would be cheaper and perhaps more productive to re-examine plant-based remedies described in ancient literature [10]. Terrestrial plants have been used as medicines in Egypt, China, India and Greece from ancient times and an impressive number of modern drugs have been developed from them [11]. The Indian subcontinent has a rich culture and there has been widespread use of medicinal herbs and spices because of their high potential abilities as traditional medicine, known worldwide as Ayurvedic, Unani and Siddha systems of medicine. The traditional communities practising for thousands of years have built a precious knowledge base about the use of the rich bio-resources of herbal remedies. The Chinese Materia Medica, which describes more than 600 medicinal plants, has documented the first record dating from about 1100 BC [12]. Documentation of the Ayurvedic system recorded in Susruta and Charaka dates back to about 1000 BC. The Greeks also contributed substantially to the rational development of herbal drugs. Dioscorides, the Greek physician (100 AD), described more than 600 medicinal plants in his work, De Materia Medica [11].

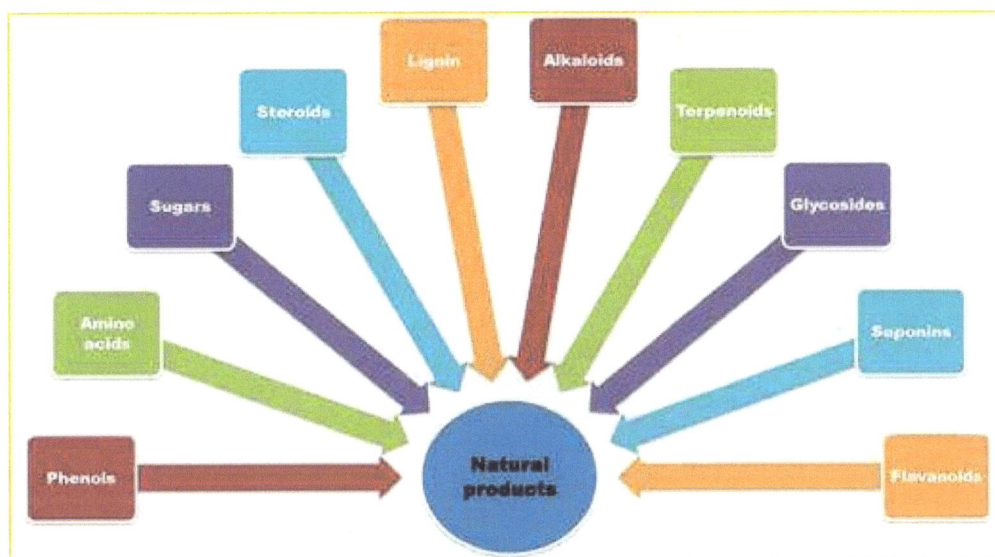

Fig. (4). Presence of different types of phytochemicals or secondary metabolites in natural products.

The exciting discoveries of many biologically active compounds/phytochemicals from natural products have encouraged the scientific community to rely on herbal medicines. Most of the bioactive molecules are part of routinely used traditional medicines and hence, tolerance and safety are relatively better known than any other chemical entities that are new for human use [13]. Bioactive compounds in plants can be defined as secondary plant metabolites eliciting pharmacological or toxicological effects in man and animals. Thus, traditional medicine-based bioactive compounds offer variety as promising new leads [14].

Several commonly and routinely used medicinal plants and spices have been used as folk medicine in the treatment of various human ailments, like *Rauwolfia serpentina*, used for centuries for its sedative effects [15], *Zingiber officinale* for digestive disorders, *Curcumin longa* as an anticancer agent, *Cinnamomum tamala* for lowering blood pressure, *Nigella sativa* as carminative, stimulant, analgesic, and anti-inflammatory agent, *Azardiracha indica* as hypoglycaemic agent [16], *Mucuna pruriens* for infertility (mucuna paper), *Withania somnifera* as an aphrodisiac, *Commiphora mukul* as hypolipidemic (mucuna paper),*Bacopa monniera* as a memory enhancer, *Pterocarpus marsupium* as an antidiabetic agent, *Catharanthus roseus* as an antidiabetic agent, *Dysoxylum* for the treatment of facial distortion in children, *Emblica officinalis* to boost defence system *and Solanum nigrum* as an anti-inflammatory agent [16].

A large number of promising novel molecules have come out of the Unani and Ayurvedic experimental database, including *Rauwolfia* alkaloids for hypertension, Psoralens in Vitiligo, *Holarrhena* alkaloids in Ameobiasis, Guggulsterons as hypolipidemic agents, *Mucuna pruriens* for Parkinson's disease, Piperdine as bioavailability enhancers, Baccosides in mental retention, Picrosides in hepatic protection, Phyllanthins as antiviral, Curcumines in inflammation, and Withanolide and many other steroidal lactones and glycosides as immunomodulators [17]. Moreover, atropine isolated from *Atropa belladonna* is used as anticholinergic, *Penicillium* sp. gives the antibiotic penicillin. Digitalin, digoxin, and digitoxin are antiarrhythmic agents obtained from *Digitalis purpurea,* whereas *Cinchona ledgeriana* produces antimalarial quinidine. These drugs have been the cornerstone for the treatment of many diseases in medical science. Furthermore, discoveries of Aspirin (a potent pain killer) from *Salix alba*, Artemisinin (an antimalarial) from *Artemesia annua*, Morphine (an opioid antagonist) from *Papaver somniferum*, Anticancer lignans from *Podophyllum peltatum*, Camptothecin (an anti-cancer agent) from *Camptotheca acuminate*, Taxol (an antineoplastic agent) from *Taxus brevifolia,* have boosted scientific community to focus on medicinal plants as phytochemical reservoirs which have been used as folk medicines for lead phytochemicals (Fig. **5**) [16].

India has about 45,000 plant species and among them, several thousand have claimed to possess medicinal properties. Recent data suggests that 80% of drug molecules are either derived as natural products or natural compounds [18]. Studies on sources of new drugs, from 1981 to 2007, revealed that almost half of the drugs approved since 1994 are based on natural products [19]. About 60% of the anticancer and 75% of anti-infective drugs approved between 1981-2002, could be traced to natural origin [20]. Though the natural products have accounted for rigorous opportunities to pharmaceutical industries by offering bioactive compounds, extracts/fractions of desired parts of natural products, and different pharmacological formulations, there are some crucial challenges which must be tackled by the medicinal chemists, such as isolation and characterization of pure active compounds in a sufficient amount, estimation of heavy metal burden in herbal formulations, *etc*.

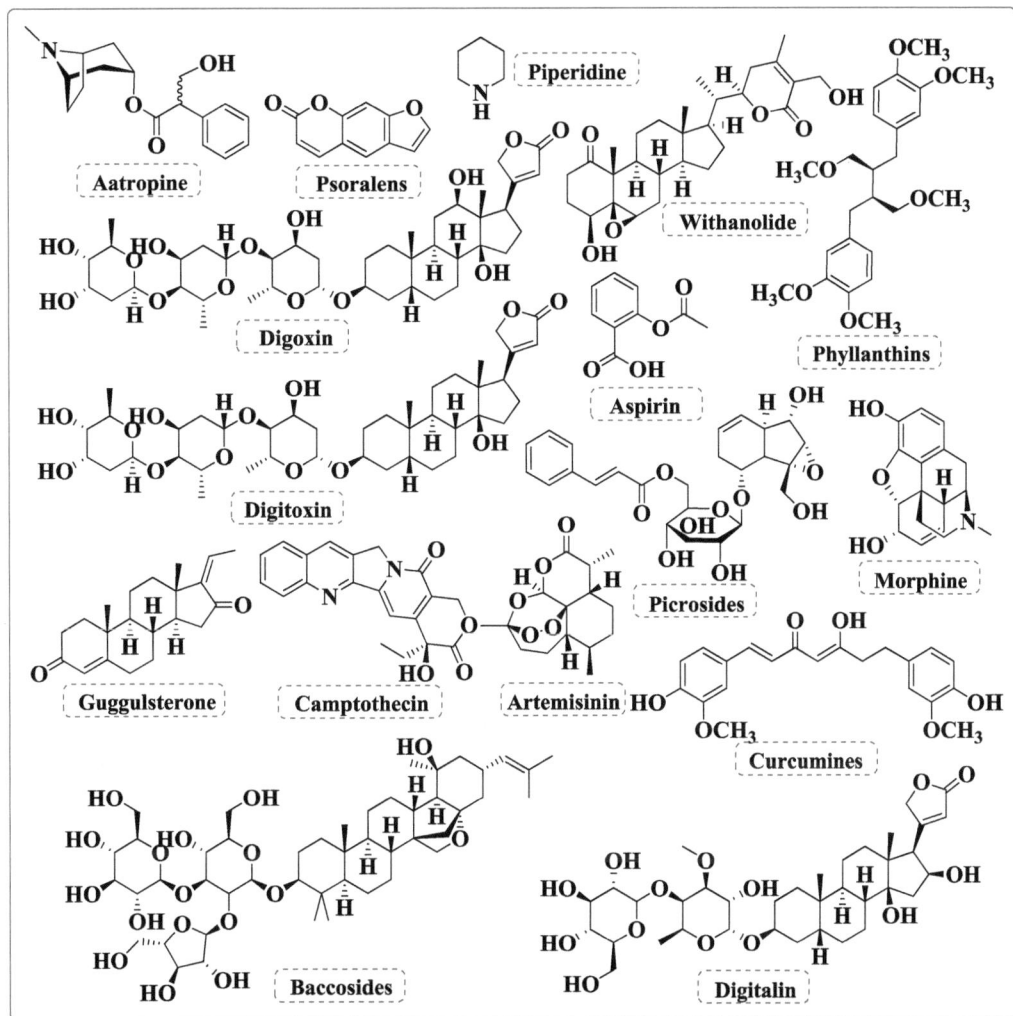

Fig. (5). Chemical structure of some precious compounds isolated from medicinal plants.

MICROBIAL METABOLITES

Recent advances in metabolomics and genome mining have uncovered a rarely understood metabolome that originates exclusively or in part from bacterial enzyme sources. Microbe oriented metabolites have made an unusual contribution to human health and well-being throughout the world. In addition to producing many primary metabolites, such as amino acids, vitamins and nucleotides, microbes can make secondary metabolites, which constitute a major part of pharmaceuticals in the market today and provide many essential products. In recent decades, the chemical diversity of microbial metabolites has facilitated them to be used as a rich source of important naturally occurring therapeutic drug leads. Microbes can produce a multitude of structurally complex metabolites by adapting processes of primary metabolism for self-defence and cell signalling purposes. Even though compounds with new structures continue to be uncovered regularly [21, 22], there is a comprehensive belief that natural products are a waning source of new leads. Back in 1928, Alexander Fleming started the microbial drug era when he discovered in a petri dish seeded with *Staphylococcus aureus* bacterium, that a compound produced by a mold killed the bacteria [23]. Later, the mold which was identified as *Penicillium notatum*, produced active agent penicillin. Penicillin was isolated as a yellow powder and used as a potent antibacterial compound during World War II. Microbial inhabitants, whether living on exposed surfaces or within our gastrointestinal tract, may produce a surprisingly different set of natural products and small molecule metabolites; these metabolites have a beneficial impact on human health and disease. A metabolite called trimethylamine N-oxide, which is a gut microbe-derived metabolite, has been causally linked to the development of CVD. Moreover, recent studies also reveal that drugging this pathway can inhibit the development of atherosclerosis in mice [24]. In recent decades, approaches such as synthetic-biology tools, system-biology guided metabolic engineering techniques and enzymatic modifications have been utilized for maximizing the applications of microbe-derived secondary metabolites [25 - 28]. Furthermore, the efficient use of microbial metabolites can be refined with considerable efforts on precise screening, higher production, and defined structural variations based on structure-activity-relationship [29, 30]. The basic methodology of attaining bioactive metabolites from a microbial source involves pure culture or single isolate method (Fig. **6**) *i.e.*:

1. Isolation and characterization of microbes.
2. Establishing culture condition for microbes and production of target metabolites.
3. Isolation, determination of structure, and assessment of activity of isolated metabolites.

The culture-based approaches explore microbial physiology and genetics in-depth. However, many microbial species cannot be cultured due to their incompatibility to grow in common media. Therefore, it is important to select and provide a suitable growth medium for a good culture of microorganisms. Also, improper optimal temperature, pH, humidity, salinity, *etc.* disturb the propagation of microbial species in the selected medium. In conclusion, the drug discovery process has experienced a paradigm shift in microbial metabolites and the comprehensive approach of mining bioactive metabolites [30 - 32].

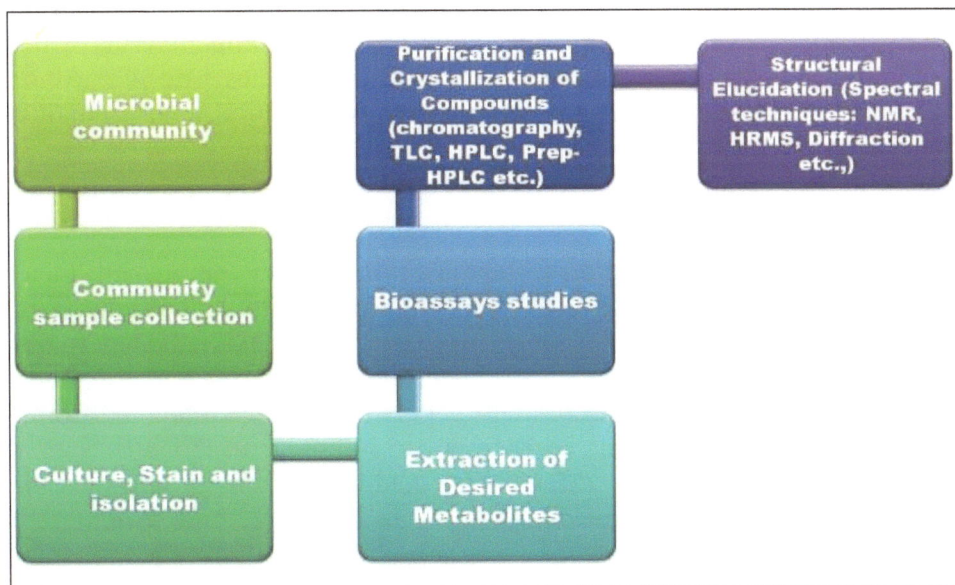

Fig. (6). Culture, staining and isolation of microbial secondary metabolites from microorganisms.

MARINE INVERTEBRATES DERIVED COMPOUNDS

Marine invertebrates are a diverse group of invertebrates that live in aquatic ecosystems, ranging from the intertidal zone to the deep-sea environments. Several marine invertebrates-based molecules have shown significant biological activities and also reported inhibitory properties in the pathogenesis of diseases. Marine invertebrate-oriented isolated compounds have been found to be pharmacologically active and are helpful in drug discovery for several deadly diseases like cancer, Acquired Immunodeficiency Syndrome (AIDS), osteoporosis, *etc.*

Historically, the dietary and medicinal uses of marine invertebrates have been described in literature across the world [33 - 35]. Hippocrates, the father of modern medicine, has described notably the use of a variety of marine invertebrates, their ingredients, and their curative effects on human health [36 - 38].

The marine world has incredible biodiversity which offers precious biologically active compounds like sterols, proteins, polysaccharides, antioxidants, pigments, *etc*. (Fig. **7,** Table **1**). Since a lot of marine invertebrates live in complex habitats and face extreme conditions, they adapt to the new environmental surroundings and produce a wide variety of different types of secondary metabolites, which cannot be found in other organisms. These secondary metabolites have been shown to be biologically active against several unmet clinical conditions [39, 40]. In recent decades, several biologically active chemical compounds have been extracted from sponges, tunicates, bryozoans, and molluscs [41, 42]. Marine invertebrates [43] can be categorized into the following major phyla, containing several species:

PORIFERA

Sponges from the phylum Porifera are the most primitive, evolutionarily, ancient metazoan animals. They are one of the richest animals with natural compounds [44 - 46]. Almost fifteen thousand sponges have been identified and many of them are found in marine water. Sponges have been grouped into three major classes: Calcarea, Demospongiae, and Hexactinellida [47]. Cytarabine, obtained from *C. crypta,* has been used for treating leukaemia and lymphoma [48 - 50]. Sponge-bacteria or fungal associations offer acetic acid-butyl-ester as antimicrobial, chloriolin B as antitumor, sorbicillactone A as anti-HIV, and roridin A as antileukemic bioactive compounds [51]. These findings emphasize the importance of sponges as a source of bioactive molecules.

CNIDARIA

Jellyfish, corals, and sea anemones belong to the phylum Cnidaria. Corals are used as an active ingredient in traditional medicine for treating various diseases, such as pulmonary tuberculosis, asthma, chronic bronchitis, urinary diseases, and cancer [33]. Corals are rich in calcium with a minor amount of magnesium and iron. Jellyfish is rich in minerals such as Na, Ca, K, and Mg [52]. Moreover, the nutrients present in jellyfish make it a low-fat and cholesterol-free marine food material. Jellyfish contains collagen as the main protein which acts as a major connective and building component of tissues, cartilages, and bones.

MOLLUSCA

The phylum Mollusca, with 50,000 living organisms, is the second-largest animal phylum on earth, with estimated 100,000–200,000 animals [53, 54]. Corn shell is a carnivorous snail belonging to the world's largest genus of marine invertebrates (Conus) under the phylum of molluscs [55]. Corn shell species secrete venom, which contains corn shell toxins- conotoxins, of bioactive peptides called conopeptides [56]. These conopeptides are widely used in clinical as well as in biomedical research for developing analgesic, anticancer, and cardio- and neuro-protective drugs [57 - 59]. Dolastatins, isolated from *Dolabella Auricularia,* are cytotoxic peptides that have entered phase II clinical trials as anticancer agents [48].

ARTHROPODA

Arthropoda is the largest phylum of living organisms. Arthropoda members have a characteristic exoskeleton and segmented bodies. Among arthropods, phylum class Crustacea is the largest and the most important group from an economic perspective. Crabs, prawns, and shrimps are some of the most attractive species due to their nutritive value and secondary metabolites [60]. Chitin is a polysaccharide primarily found in the exoskeleton of crustaceans. It is the main raw material of chitosan, which has a wide range of applications as dietary supplements, functional foods, drugs, cosmetics, antioxidants, and immune stimulants. Chitin and chitosan show hemostasis, drug release, antitumor, anticoagulant, wound healing, antiulcer, antimicrobial, and immunomodulatory properties [61]. Several antimicrobial peptides have been isolated from crustaceans and evaluated for their ability to control pathogenic microorganisms. Over about 50 different sequences of crustin type antimicrobial peptides have been isolated from crabs, crayfish, and white shrimp.

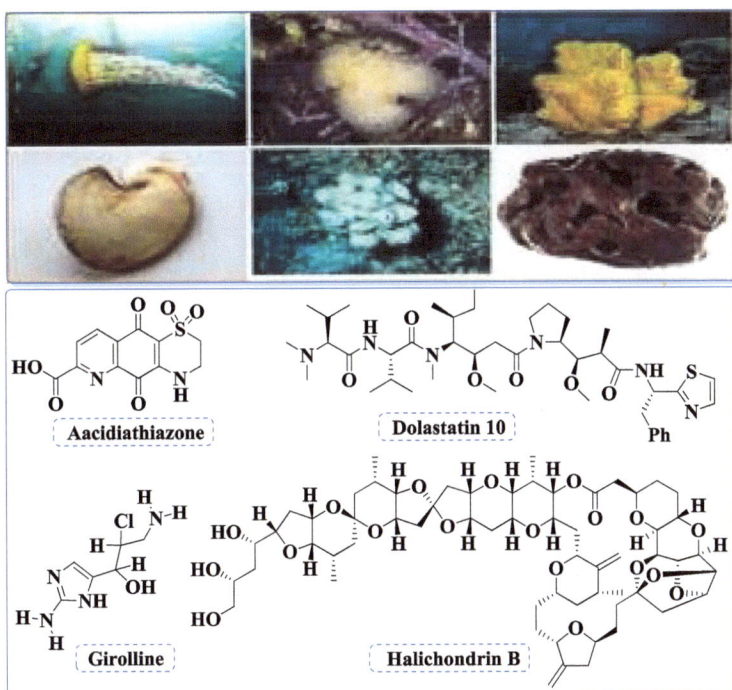

Fig. (7). Marine invertebrates and bioactive isolated molecules from marine species.

Table 1. List of some marine invertebrate derived biologically active natural products/compounds.

Phylum	Family	Species	Compound Isolated	Biological Activity
Porifera	Halichondriidae	*Stylotella aurantium* Kelly-Borges &Bergquist, 1988	Debromohymenialdisine	Anticancer, Alzheimer's disease
	Halichondriidae	*Halichondria okadai* Kadota, 1922	Halichondrins	Anticancer
	Halichondriidae	*Pseudaxinyssacantharella* Lévi, 1983	Girolline	Antimalaria
	Petrosiidae	*Petrosiacontignata* Thiele, 1899	Contignasterol	Against asthma and hemodynamic disorders
	Theonellidae	*Discodermiadissoluta* Schmidt, 1880	(+)-Discodermolide	Anticancer
	Plakinidae	Sponge (*P. angulospiculatus*)	Plakortide P	Antineuroinflammatory
Cnidaria	Gorgoniidae	*Pseudopterogorgiaelisabethae* Bayer 1961	Pseudopterosins	Dermal infection and wound healing
	Alcyoniidae	*Lobophytum crissum*	Cembranolides	Inhibitors of COX-2
	Ellisellidae	*Junceella fragilis*	Frajunolides	Anti-inflammatory action in human neutrophils
Mollusca	Aplysiidae	*Dolabella auricularia* Lightfoot, 1786	Dolastatin 10	Anticancer
	Plakobranchidae	*Elysiarufescens* Pease, 1871	Kahalalide F	Against cancer and skin conditions
	Mactridae	*Spisulapolynyma* Stimpson, 1860	Spisulosine	Anticancer
Bryozoa	Bugulidae	*Bugulaneritina* Linnaeus, 1758	Bryostatin 1	Anticancer, Alzheimer's disease
Tunicata	Didemnidae	*Lissoclinum sp.*	Mandelalides	Anticancer
	Perophoridae	*Ecteinascidia turbinata* Herdman, 1880	Trabectedin	Anticancer
	Polyclinidae	*Aplidiumalbicans* Milne-Edwards, 1841	Plitidepsin	Anticancer
Hemichordata	Cephalodiscidae	*Cephalodiscusgilchristi* Ridewood, 1908	Cephalostatin 1	Anticancer
Chordata	Ascidian	*Ascidian Aplidium*	Alkaloids Ascidiathiazone	Anti-inflammatory action in human neutrophils
		Ascidian (*Synoicum* sp.)	Rubrolide O	Anti-inflammatory action in human neutrophils

RECOMBINANT DNA TECHNOLOGY

Recombinant DNA technology or genetic engineering is a modern concept for the synthesis of therapeutic agents. These therapeutics agents are commonly known as biologics/ bioengineered, biopharmaceuticals, recombinant DNA-expressed products, or genetically engineered drugs. There are various recombinant technology-based drugs available, including erythropoietin, coagulation modulators, interferons, interleukins, granulocyte colony-stimulating factors, anti-rheumatoid drugs, and other agents like TNF, becaplermin, hepatitis-B vaccine antibodies, *etc*. The recombinant DNA technology procedure starts from the identification of genes responsible for the desired products. Firstly, the gene is isolated from human cells and inserted into another carrier, known as vector cells, such as bacteria (*Escherichia coli*) or yeast (*Saccharomyces cerevisiae, Hansenulla polymorpha, Pichia pastoris*). This leads to the proliferation and production of a sufficient quantity of desired products (Fig. **8**). Various enzymes have an important role in recombinant DNA technology, such as DNA polymerases, ligases, kinases, alkaline phosphatases, and nucleases [62].

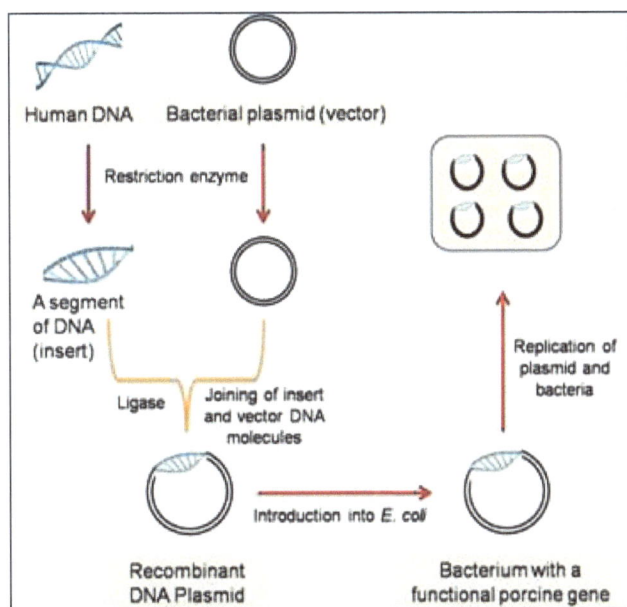

Fig. (8). Recombinant DNA technique.

ANIMAL-BASED MEDICINES

Animals have been traditionally used for the treatment of many human illnesses and diseases. Animals have been identified as a potent medicinal reservoir of therapeutics. The healing/treatment therapy for different human ailments by using animal therapeutics is known as **'Zootherapy'**. The use of an animal's body parts as medicine is relevant because it implies additional pressure over critical wild populations. In India, about 15-20% of the Ayurvedic medicines are animal-derived substances [63] and approximately 380 types of animal substances mentioned in ancient literature in Charaka Samhita have been widely used for different clinical conditions.

Animal-based medicines have been derived from different parts of the animal body and metabolism products. These remedies are administered to the patient in the form of plasters, decoction, reek, and food. Respiratory diseases, asthma, bronchitis, stroke, and wounds are the most usual clinical conditions met by these animal-oriented medicines, which may include scales, spur, shell, fat, skin, globe of the eye, tentacles, and otolith. Also, insects have been considered important sources of drugs due to their immunological, analgesic, antibacterial, diuretic, anaesthetic, and anti-rheumatic properties [64]. For example, tetrameric polypeptide melittin, a major component of bee venom, has shown anti-arthritic and anti-inflammatory effects [65]. Margatoxin, obtained from the venom of scorpion *Centruroides margaritatus,* blocks lymphocytes activation and the production of interleukin-2 by human T- lymphocytes.

DESIGNING OF NEW CHEMICAL ENTITIES (NCES)

New Chemical Entities (NCEs) are drugs that contain no active moiety which has previously been approved as a drug by concerned authorities. The development of NCEs requires significant investment in many ways. NCEs are the molecules developed by innovators in the early phase of drug discovery/development which after passing different stages undergo clinical trials. Small molecules-based drugs represent the majority of new chemical entities (NCEs). The synthesis or isolation of an NCE from natural sources is the first step in the process to convert it into NDEs. This early phase of drug discovery and development is technically challenging and uncertain. There are various issues such as raw materials, yield, physical and chemical properties, stereochemistry, functional groups, framework polymorphism, *etc.* which must be kept in mind before preparing a scheme to synthesize or design the NCEs.

Organic synthesis of NCEs demands the design and construction of a novel, biologically active, safe, and suitable target-oriented compound. Additionally, the flexibility of structural optimization/improvement is required which is necessary for safety and clinical studies. Small chiral synthetic molecules with multiple activities are preferred as oral drugs. The synthesized NCEs should not only be biologically active but also must have suitability for administration, should be non-toxic, and should possess excellent pharmacokinetic properties.

Designing chiral, optically active chemical compounds and stereo-selective transformations is an important part of designing New Drug Entities. Chiral, optically active drugs are occupying 68% of the top 200 branded drugs and 62.5% of the top 200 generic drugs [66]. Designing of NCEs requires a medicinal chemist to use scientific skills and experience to envisage a compound and then employ appropriate reactions to achieve the desired NCE. In recent advancements, organic chemists have developed a methodology to design NCEs to activate carbon-hydrogen bonds (C-H) and fluorination of complicated organic molecules of beneficial interest. Organocatalytic synthetic reactions selectively facilitate the synthesis of enantiomers which have shown promising biological activity [67].

The latest developments in organic chemistry have helped medicinal chemists to assemble the NCEs through target-based, structure-based, ligand-based, and phenotype-based approaches, which increases the probability of NCEs converting into NDEs. The recent developments in natural products have emerged as alternative treatment strategies for many clinical conditions where any specific drug molecule is not available (detailed discussion in subsequent heading). For example, in the very recent outbreak of the novel coronavirus (SARS-CoV-2 also known as COVID-19), there is no specific treatment or antiviral drug available, although natural products, such as *Withania somnifera, Tinospora cordifolia, Ocimum sanctum, Zingiber officinale, cinnamon, black pepper, Nigella sativa,* and other natural immunity boosters, act as an alternative strategy and have played a very significant role in the management of COVID-19 patients and have reduced the mortality burden to a remarkable extent.

With the emergence of computational biology, bioinformatics and molecular docking have now become an indispensable tool in structural biology and computer-aided drug designing (CADD) and are very helpful in the prediction of the probable activity spectrum of a probable drug compound, simultaneously providing insight about various ADMET (Absorption, Distribution, Metabolism, Excretion, and Toxicological) aspects, which are the most important part for safe and effective drugs [68]. To balance the computational efficiency and accuracy, a hieratical strategy employing different types of scoring functions is applied in both the drug lead identification and optimization phases [69]. This *in silico* drug discovery process involves;

- Generation of small compounds library for the testing and screening.
- Interaction of selected molecules/ligands with 3D models of protein binding site (PBD)/ receptors by molecular docking to find out the hits.
- Pharmacokinetic and toxicological studies of hits.

Computer-Aided Drug Design (CADD) tools are credited for the discovery of numerous approved drugs in the field of pharmaceutical industries, including Captopril (an angiotensin-converting enzyme (ACE) inhibitor for the treatment of hypertension [70]), Dorzolamide (a carbonic anhydrase (AC) inhibitor for cystoid macular oedema, glaucoma) [71, 72], the protease inhibitors, and the antivirals-Ritonavir, Indinavir and Saquinavir, the most potent anti-HIV drugs [73, 74].

DEGRADATION OF DRUGS

Medicines have an important role in human health and the prevention of diseases in both humans and animals. However, these medicines are released into the environment in different ways (Fig. **9**). Wide ranges of human and veterinary therapeutics such as antibiotics, pain relievers, supplements, *etc.* are produced and used in a very huge amount and are released into the environment by various routes. Recent monitoring has detected low levels of a range of pharmaceuticals, including hormones, steroids, antibiotics, and parasiticides, in soils, surface water, and groundwater [75, 76]. There are many ways (given below) that medications, at varying concentration levels, can enter the environment.

• Drug residues/wastes/by-products/effluents can enter water reservoirs, air, soils, landfills, *etc.* in the manufacturing process.
• After administration, medicines are absorbed, metabolized, and then excreted in trace amounts into the sewer system, and finally distributed in the environment.
• Veterinary pharmaceuticals used on farm animals are excreted into soils and water surfaces.
• Unused/expired medicines are often disposed of into the water supply through sinks, toilets, and landfills.

It is well established that 'drug residues' can impact microorganisms. The emergence of antibiotic-resistant pathogens clearly shows this point well. This resistance in bacteria/pathogens is most common in places where antibiotics are heavily used.

The only way to redress the current practices is to pinpoint how the environment is vulnerable to pharmaceutical exposure. A multifaceted strategy is necessary at the level of society, including consumer education on how to properly dispose of medications, safer production practices at pharmaceutical units, and national legislation to govern the drug manufacturing and disposal process. Today, the movement to protect the environment and waterways from pharmaceutical manufacturing practices is known as 'green pharmacy'.

The 'Pharmaceutical Residue' in the environment can be removed through physical processes, such as sorption/ volatilization, chemical reaction, biological degradation, treatment with ozone, *etc*. For example, the antibiotic ciprofloxacin can be removed by strong sorption onto suspended solids of sewage sludge. The anti-inflammatory drug, diclofenac, undergoes significant biodegradation in aged, activated sludge.

There seems to be a dearth of research in the critical area of the environmental impact of drugs and there must be an emergent concern about how to design more benign drugs. One very important strategy is to make or construct drugs that are more environmentally friendly. Human excretion of residues of drugs is a potent source of water contamination, and thus eco-friendly and readily biodegradable drug design will be savvier and healthier. Drug companies should strive to make more eco-friendly versions of their drugs. Although drug manufacturers may be the ones to drive major change, the government agencies and consumers' contribution is also very important.

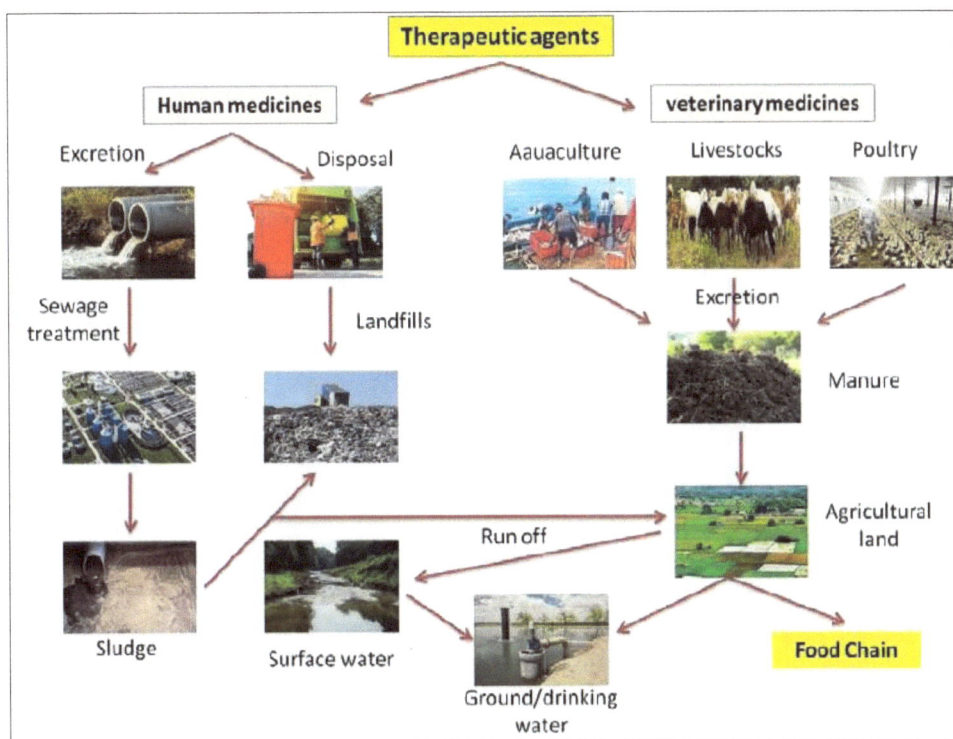

Fig. (9). Environmental degradation of drugs.

CHALLENGES IN MEDICINAL CHEMISTRY

Medicinal chemistry has made tremendous transformations and offered boundless drugs molecules for almost every clinical condition. This has been accomplished through several technological innovations and advancements in recent decades. Despite several innovations in medicinal chemistry, there are so many challenges yet unwrapped in developing potent NDEs. For example, several challenges associated with the construction of NCEs. The key reason is uncertainty. As with anything new, the pathway to manufacturing is completely unchartered territory. No one can be assured about the new chemical entity as to how the new molecule will react under different conditions, and how long it may take to achieve the desired outcome. The complexity arises in terms of physical and chemical properties, such as stereochemistry, functional group orientation, and molecular skeleton, in addition to biological activity. Medicinal chemists are as yet mainly considering 'how to synthesize' rather than 'what to synthesize' to cure a particular disease.

One of the greatest challenges in medicinal chemistry is the failure of NCEs in pre-clinical stages. Reducing the number of these NCEs/drugs that fail in pre-clinical stages is much needed since this is responsible for the very high cost of bringing a drug to the market (Fig. **10**). The accountant's mantra of 'fail early/fail cheap' is now central to the drug discovery process. The biological aspects such as ADMET properties, particularly toxicology, such as irreversible protein binding, idiosyncratic toxicity, mutagenic toxicity *etc.* largely decide whether a drug molecule is rejected immediately or taken forward for optimisation. For example, numerous medicines have been registered or approved for clinical trials for COVID-19, but the competency of each drug is quite different and still unknown, moreover, many of them have failed due to toxicology and adverse effects at different clinical trial phases. However, the Russian vaccine 'Sputnik V' is the first registered vaccine for COVID-19 and is now being used for this disease.

The ADMET challenge in medicinal chemistry is very fundamental and each NDE must clear this test. It is quite evident that ADMET profiling, using *in silico* and cellular approaches, has successfully resulted in a reduced attrition rate. Medicinal chemists have developed various *in silico* models and mathematical tools, along with *in vitro* experimentation to assess the potential toxicology problems and reducing these problems to make effective clinical candidates. These approaches are employed by medicinal chemists and pharmaceutical industries to developed capable NDE.

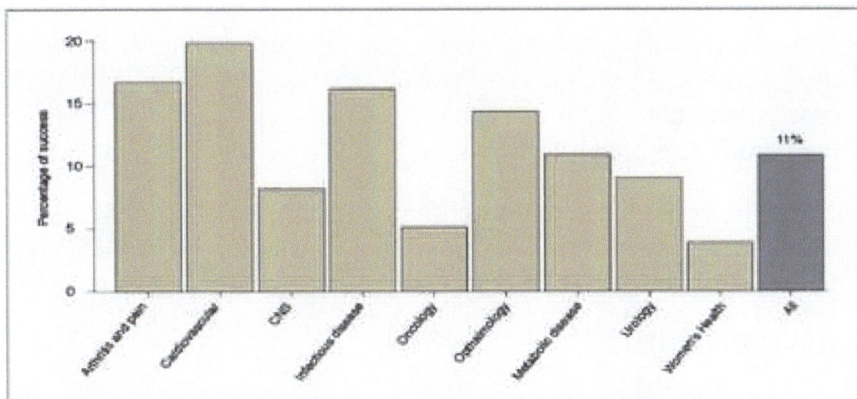

Fig. (10). Percentage of success (Attrition rate) of NCEs) (Image Courtesy: https://www.ddw-online.com/medicinal-chemistry-progress-through-innovation-779-200608/).

In this rapidly changing and challenging globalization scenario, medicinal chemistry must play a key role in drug designing and discovery, through lead discovery and optimization, to check all the essential needs for drug candidates to become new medicines. However, conventional tools for medicinal chemistry are facing serious challenges. Medicinal chemistry is deeply challenged by the designing and creation of novel, biologically active, safe, and targeted molecules with scale-up of compound quantities for safety and clinical studies. The complexity of marketing and the high attrition rate is the frequent failure and high risk in searching for new drug entities [77]. Therefore, a key goal is to reduce this high attrition rate by transforming drug discovery into a high-throughput, rational process. The Active Pharmaceutical Ingredient (API) is also very essential due to safety studies (toxicology) and the optimization of suitable dosage forms and testing in clinical phases I–III studies.

Lead generation and lead optimization involve the chemical synthesis of novel molecules, which fulfil all the desired criteria of drug discovery. The challenges of a high-quality lead compound are [78, 79]:

• Its synthetic tractability
• The patentability of the series around the lead
• Availability of chemistry space for optimization
• Acceptable solubility, permeability, and protein binding
• Lack of inhibition of cytochrome P450 (CYP; family of enzymes responsible for oxidative drug metabolism)
• Lack of inhibition of human ether-a`-go-go related gene (hERG; a gene that codes for a potassium ion channel in the heart, inhibition of which can cause

cardiac arrest)
- Acceptable absorption–distribution–metabolism–excretion (ADME)
- Its confirmed interaction with the biological target (*e.g.*, by X-ray or NMR)
- Defined apparent structure-activity relationship (SAR)
- Its confirmed biological activity *in vivo*.

There are other challenges, such as sourcing raw materials, cost, yields, analytical problems, regulatory parameters, financial and social issues which are also very important and must be considered in drug development and discovery.

CONCLUDING REMARKS

Based on the above discussion, it can be concluded that drug discovery is a key discipline of medicinal chemistry and is fundamentally important to the pharmaceutical industries. The different key sources of drug molecules discussed above have accounted for unvalued lead drugs to meet the clinical needs. Although in contrast to synthetic and medicinal plant-based classical drug discovery, the recent exploration of microbial, marine, recombinant DNA, and genetic engineering-oriented leads has stimulated the researchers for the discovery of NDEs. Contemporary drug discovery and design still face many scientific, social, and economic challenges to offer lead compounds competently. Moreover, the large delays and huge expenses in the conventional method of drug discovery have prompted the scientific community to focus on strategic approaches to solve the issues in lead candidate identification and optimization, through high throughput screening, bioinformatics, computer-aided drug design, and virtual screening. Despite the distinctive achievements in medicinal chemistry, there are several unmet clinical needs, including COVID-19, for which no lead drug or vaccine is available for the treatment, except some supportive medicines. This has driven the medicinal chemist to explore the innovations in synthetic methodologies and reactions, including microwave chemistry, continuous flow, and automated synthesis, and the development of modular, convergent synthetic reactions. Also, the 'drug repositioning' may be an optional approach to traditional drug discovery and development. Furthermore, deploying 'Artificial Intelligence' to discover NDEs will accelerate the drug development process, with automated compound/drug optimization. Overall, the main objective of the medicinal chemists is to make the drug discovery process more efficient and less time-consuming, for the optimization of lead compounds into safe and efficient NDEs. The successful development of NDEs requires interdisciplinary teamwork as no single approach can be a panacea. We hope this perspective will contribute towards overcoming the problems with the drug development process in medicinal chemistry.

CONSENT FOR PUBLICATION

Not Applicable.

CONFLICT OF INTEREST

The author confirms that this chapter contents have no conflict of interest.

ACKNOWLEDGEMENT

Declared none.

REFERENCES

[1]   Wermuth CG, Ganellin CR, Lindberg P, Mitscher LA. Glossary of terms used in medicinal chemistry (IUPAC Recommendations 1998). Pure Appl Chem 1998; 70(5): 1129-43.
        [http://dx.doi.org/10.1351/pac199870051129]

[2]   Wadud A, Prasad PV, Rao MM, Narayana A. Evolution of drug: a historical perspective. Bull Indian Inst Hist Med Hyderabad 2007; 37(1): 69-80.
        [PMID: 19569453]

[3]   Imming P. Medicinal Chemistry: definitions and objectives, drug activity phases, drug classification systems.The Practice of Medicinal Chemistry. Academic Press 2008; pp. 63-72.
        [http://dx.doi.org/10.1016/B978-0-12-374194-3.00002-0]

[4]   Barends D, Dressman J, Hubbard J, *et al.* Multisource (Generic) Pharmaceutical Products: Guidelines on Registration Requirements to Establish Interchangeability Draft Revision. Geneva, Switzerland: WHO 2005.

[5]   Ganesan A, Proudfoot J. Analogue-based drug discovery. Wiley-VCH 2010.

[6]   Šunjić V, Parnham MJ. Organic Synthesis in Drug Discovery and Development.Signposts to Chiral Drugs. Basel: Springer 2011; pp. 1-12.
        [http://dx.doi.org/10.1007/978-3-0348-0125-6_1]

[7]   Lipinski CA, Lombardo F, Dominy BW, Feeney PJ. Experimental and computational approaches to estimate solubility and permeability in drug discovery and development settings. Adv Drug Deliv Rev 2001; 46(1-3): 3-26.
        [http://dx.doi.org/10.1016/S0169-409X(00)00129-0] [PMID: 11259830]

[8]   Veber DF, Johnson SR, Cheng HY, Smith BR, Ward KW, Kopple KD. Molecular properties that influence the oral bioavailability of drug candidates. J Med Chem 2002; 45(12): 2615-23.
        [http://dx.doi.org/10.1021/jm020017n] [PMID: 12036371]

[9]   Akerele O. Medicinal plants and primary health care: an agenda for action. Fitoterapia 1988; 59: 355-63.

[10]   Holland BK. Prospecting for drugs in ancient texts. Nature 1994; 369(6483): 702-, 369, 702.
        [http://dx.doi.org/10.1038/369702a0] [PMID: 8008059]

[11]   Shoeb M. Anti-cancer agents from medicinal plants. Bang J of Pharmacol 2006; 1: 35-41.

[13]   Patwardhan B, Vaidya ADB, Chloghade M. Ayurveda and natural products drug discovery. Curr Sci 2004; 86: 10.

[12]   Cragg GM, Newman DJ, Weiss RB. Coral reefs, forests, and thermal vents: the worldwide exploration of nature for novel antitumor agents. 1997.

[14]   Koehn FE, Carter GT. The Evolving role of Natural Products in Drug Discovery 2005.

[http://dx.doi.org/10.1038/nrd1657]

[15] Jin-Ming K. Ngoh-Khang, G.O.H., Lian-Sail, G.O.H. Fatt, C. Recent advances in traditional plant drugs and orchids. Acta Pharmacol Sin 2003; 24: 7-21.

[16] Mahdi AA, Ansari JA, Khan HJ, Fatima N, Lakshmi V, Ahmad MK. Khan, A.R. Anticancerous Medicinal Plants: A Review. Int. J of Adv. in Pharma. Res 2013; 13: 1706-22.

[17] Patwardhan B. Ayurveda: The 'Designer' medicine: A review of Ethnopharmacology and Bioprospecting Research. IDrugs 2000; 37: 2000.

[18] Harvey AL. Natural products in drug discovery. Drug Discov Today 2008; 13(19-20): 894-901.
[http://dx.doi.org/10.1016/j.drudis.2008.07.004] [PMID: 18691670]

[19] Butler MS. Natural products to drugs: natural product-derived compounds in clinical trials. Nat Prod Rep 2008; 25(3): 475-516.
[http://dx.doi.org/10.1039/b514294f] [PMID: 18497896]

[20] Gupta R, Gabrielsen B, Ferguson SM. Nature's medicines: traditional knowledge and intellectual property management. Case studies from the National Institutes of Health (NIH), USA. Curr Drug Discov Technol 2005; 2(4): 203-19.
[http://dx.doi.org/10.2174/1570163057752029937] [PMID: 16475917]

[21] Cragg GM, Boyd MR, Khanna R, Newman DJ, Sausville EA. Natural product drug discovery and development.Phytochemicals in Human Health Protection, Nutrition, and Plant Defense. Boston, MA: Springer 1999; pp. 1-29.
[http://dx.doi.org/10.1007/978-1-4615-4689-4_1]

[22] Newman DJ, Cragg GM, Snader KM. Natural products as sources of new drugs over the period 1981-2002. J Nat Prod 2003; 66(7): 1022-37.
[http://dx.doi.org/10.1021/np030096l] [PMID: 12880330]

[23] Fleming A. On the antibacterial action of cultures of a penicillium, with special reference to their use in the isolation of B. influenzae. Br J Exp Pathol 1929; 10: 226.

[24] Li Z, Wu Z, Yan J, *et al.* Gut microbe-derived metabolite trimethylamine N-oxide induces cardiac hypertrophy and fibrosis. Lab Invest 2019; 99(3): 346-57.
[http://dx.doi.org/10.1038/s41374-018-0091-y] [PMID: 30068915]

[25] Dhakal D, Le TT, Pandey RP, *et al.* Enhanced production of nargenicin A(1) and generation of novel glycosylated derivatives. Appl Biochem Biotechnol 2015; 175(6): 2934-49.
[http://dx.doi.org/10.1007/s12010-014-1472-3] [PMID: 25577346]

[26] Koju D, Maharjan S, Dhakal D, Yoo JC, Sohng JK. Effect of different biosynthetic precursors on the production of nargenicin A1 from metabolically engineered Nocardia sp. CS682. J Microbiol Biotechnol 2012; 22(8): 1127-32.
[http://dx.doi.org/10.4014/jmb.1202.02027] [PMID: 22713990]

[27] Le TT, Pandey RP, Gurung RB, Dhakal D, Sohng JK. Efficient enzymatic systems for synthesis of novel α-mangostin glycosides exhibiting antibacterial activity against Gram-positive bacteria. Appl Microbiol Biotechnol 2014; 98(20): 8527-38.
[http://dx.doi.org/10.1007/s00253-014-5947-5] [PMID: 25038930]

[28] Jha AK, Dhakal D, Van PTT, *et al.* Structural modification of herboxidiene by substrate-flexible cytochrome P450 and glycosyltransferase. Appl Microbiol Biotechnol 2015; 99(8): 3421-31.
[http://dx.doi.org/10.1007/s00253-015-6431-6] [PMID: 25666682]

[29] Dhakal D, Sohng JK. Commentary: Toward a new focus in antibiotic and drug discovery from the Streptomyces arsenal. Front Microbiol 2015; 6: 727.
[http://dx.doi.org/10.3389/fmicb.2015.00727] [PMID: 26236304]

[30] Dhakal D, Sohng JK. Coalition of biology and chemistry for ameliorating antimicrobial drug discovery. Front Microbiol 2017; 8: 734.

[http://dx.doi.org/10.3389/fmicb.2017.00734] [PMID: 28522993]

[31] Dhakal D, Dhakal Y, Sapkota D, Sohng J K. Drug discovery based on microbial metabolism and communities. EC. Microbiol., (SI), 2017, 1, 6-9.

[32] Nemergut DR, Schmidt SK, Fukami T, *et al.* Patterns and processes of microbial community assembly. Microbiol Mol Biol Rev 2013; 77(3): 342-56.
       [http://dx.doi.org/10.1128/MMBR.00051-12] [PMID: 24006468]

[33] Gopal R, Vijayakumaran M, Venkatesan R, Kathiroli S. Marine organisms in Indian medicine and their future prospects. Nat Prod Rad 2008; 7: 139-45.

[34] Lev E. Traditional healing with animals (zootherapy): medieval to present-day Levantine practice. J Ethnopharmacol 2003; 85(1): 107-18.
       [http://dx.doi.org/10.1016/S0378-8741(02)00377-X] [PMID: 12576209]

[35] Yesilada E. Past and future contributions to traditional medicine in the health care system of the Middle-East. J Ethnopharmacol 2005; 100(1-2): 135-7.
       [http://dx.doi.org/10.1016/j.jep.2005.06.003] [PMID: 15994043]

[36] Grammaticos PC, Diamantis A. Useful known and unknown views of the father of modern medicine, Hippocrates and his teacher Democritus. Hell J Nucl Med 2008; 11(1): 2-4.
       [PMID: 18392218]

[37] Voultsiadou E. Therapeutic properties and uses of marine invertebrates in the ancient Greek world and early Byzantium. J Ethnopharmacol 2010; 130(2): 237-47.
       [http://dx.doi.org/10.1016/j.jep.2010.04.041] [PMID: 20435126]

[38] Voultsiadou E, Vafidis D. Marine invertebrate diversity in Aristotle's zoology. Contrib Zool 2007; 76: 103-20.
       [http://dx.doi.org/10.1163/18759866-07602004]

[39] Rasmussen RS, Morrissey MT. Marine biotechnology for production of food ingredients. Adv Food Nutr Res 2007; 52: 237-92.
       [http://dx.doi.org/10.1016/S1043-4526(06)52005-4] [PMID: 17425947]

[40] Plaza M, Cifuentes A, Ibáñez E. In the search of new functional food ingredients from algae. Trends Food Sci Technol 2008; 19: 31-9.
       [http://dx.doi.org/10.1016/j.tifs.2007.07.012]

[41] Donia M, Hamann MT. Marine natural products and their potential applications as anti-infective agents. Lancet Infect Dis 2003; 3(6): 338-48.
       [http://dx.doi.org/10.1016/S1473-3099(03)00655-8] [PMID: 12781505]

[42] Haefner B. Drugs from the deep: marine natural products as drug candidates. Drug Discov Today 2003; 8(12): 536-44.
       [http://dx.doi.org/10.1016/S1359-6446(03)02713-2] [PMID: 12821301]

[43] Thorpe JP, Solé-Cava AM, Watts PC. Exploited marine invertebrates: genetics and fisheries.Marine genetics. Dordrecht: Springer 2000; pp. 165-84.
       [http://dx.doi.org/10.1007/978-94-017-2184-4_16]

[44] Belarbi H, Contreras Gómez A, Chisti Y, García Camacho F, Molina Grima E. Producing drugs from marine sponges. Biotechnol Adv 2003; 21(7): 585-98.
       [http://dx.doi.org/10.1016/S0734-9750(03)00100-9] [PMID: 14516872]

[45] Blunt JW, Copp BR, Hu WP, Munro MH, Northcote PT, Prinsep MR. Marine natural products. Nat Prod Rep 2007; 24(1): 31-86.
       [http://dx.doi.org/10.1039/b603047p] [PMID: 17268607]

[46] Hu GP, Yuan J, Sun L, *et al.* Statistical research on marine natural products based on data obtained between 1985 and 2008. Mar Drugs 2011; 9(4): 514-25.
       [http://dx.doi.org/10.3390/md9040514] [PMID: 21731546]

[47]   Fieseler L, Horn M, Wagner M, Hentschel U. Discovery of the novel candidate phylum "Poribacteria" in marine sponges. Appl Environ Microbiol 2004; 70(6): 3724-32.
[http://dx.doi.org/10.1128/AEM.70.6.3724-3732.2004] [PMID: 15184179]

[48]   Schwartsmann G, Brondani da Rocha A, Berlinck RG, Jimeno J. Marine organisms as a source of new anticancer agents. Lancet Oncol 2001; 2(4): 221-5.
[http://dx.doi.org/10.1016/S1470-2045(00)00292-8] [PMID: 11905767]

[49]   Sipkema D, Franssen MC, Osinga R, Tramper J, Wijffels RH. Marine sponges as pharmacy. Mar Biotechnol (NY) 2005; 7(3): 142-62.
[http://dx.doi.org/10.1007/s10126-004-0405-5] [PMID: 15776313]

[50]   Essack M, Bajic VB, Archer JA. Recently confirmed apoptosis-inducing lead compounds isolated from marine sponge of potential relevance in cancer treatment. Mar Drugs 2011; 9(9): 1580-606.
[http://dx.doi.org/10.3390/md9091580] [PMID: 22131960]

[51]   Thomas TRA, Kavlekar DP, LokaBharathi PA. Marine drugs from sponge-microbe association--a review. Mar Drugs 2010; 8(4): 1417-68.
[http://dx.doi.org/10.3390/md8041417] [PMID: 20479984]

[52]   Hsieh YHP, Leong FM, Barnes KW. Inorganic constituents in fresh and processed cannonball jellyfish (Stomolophus meleagris). J Agric Food Chem 1996; 44: 3117-9.
[http://dx.doi.org/10.1021/jf950223m]

[53]   Bouchet P, Duarte CM. The exploration of marine biodiversity: scientific and technological challenges. Fundación BBVA 2006; 33: 1-34.

[54]   Pechenic JA. Biology of Invertebrates. 4th ed., New York: McGraw Hill 2000.

[55]   Bingham JP, Mitsunaga E, Bergeron ZL. Drugs from slugs--past, present and future perspectives of ω-conotoxin research. Chem Biol Interact 2010; 183(1): 1-18.
[http://dx.doi.org/10.1016/j.cbi.2009.09.021] [PMID: 19800874]

[56]   Layer RT, McIntosh JM. Conotoxins: therapeutic potential and application. Mar Drugs 2006; 4: 119-42.
[http://dx.doi.org/10.3390/md403119]

[57]   Han Y, Huang F, Jiang H, *et al.* Purification and structural characterization of a D-amino acid-containing conopeptide, conomarphin, from Conus marmoreus. FEBS J 2008; 275(9): 1976-87.
[http://dx.doi.org/10.1111/j.1742-4658.2008.06352.x] [PMID: 18355315]

[58]   Layer RT, McIntosh JM. Conotoxins: therapeutic potential and application. Mar Drugs 2006; 4: 119-42.
[http://dx.doi.org/10.3390/md403119]

[59]   Twede VD, Miljanich G, Olivera BM, Bulaj G. Neuroprotective and cardioprotective conopeptides: an emerging class of drug leads. Curr Opin Drug Discov Devel 2009; 12(2): 231-9.
[PMID: 19333868]

[60]   Smith VJ, Fernandes JM, Kemp GD, Hauton C. Crustins: enigmatic WAP domain-containing antibacterial proteins from crustaceans. Dev Comp Immunol 2008; 32(7): 758-72.
[http://dx.doi.org/10.1016/j.dci.2007.12.002] [PMID: 18222540]

[61]   Prashanth K H, Tharanathan R N. Chitin/chitosan: modifications and their unlimited application potential—an overview. Trends. Food. Sci. Technol., 2007.18, 117-131.

[62]   Glick BR, Patten CL. Molecular biotechnology: principles and applications of recombinant DNA. John Wiley & Sons 2017; Vol. 34.
[http://dx.doi.org/10.1128/9781555819378]

[63]   Unnikrishnan PM. Animals in ayurveda. Amruth 1998; 1: 1-23.

[64]   Yamakawa M. Insect antibacterial proteins. J Sericultural Sci Jap 1998; 67: 163-82.

[65]   Bisset NG. One man's poison, another man's medicine? J Ethnopharmacol 1991; 32(1-3): 71-81.
[http://dx.doi.org/10.1016/0378-8741(91)90105-M] [PMID: 1881170]

[66]   Brooks E, Brichacek M, Mc Grath N, Morton J, Batory L, Njardarson JT. Compilation of the Njardarson Group. Cornell University 2008.

[67]   Rotella DP. 2016.The critical role of organic chemistry in drug discovery
[http://dx.doi.org/10.1021/acschemneuro.6b00280]

[68]   Meng XY, Zhang HX, Mezei M, Cui M. Molecular docking: a powerful approach for structure-based drug discovery. Curr Comput Aided Drug Des 2011; 7(2): 146-57.
[http://dx.doi.org/10.2174/157340911795677602] [PMID: 21534921]

[69]   Bohm HJ, Stahl M. The use of scoring functions in drug discovery applications. Rev Comput Chem 2002; 18: 41-88.
[http://dx.doi.org/10.1002/0471433519.ch2]

[70]   Talele TT, Khedkar SA, Rigby AC. Successful applications of computer aided drug discovery: moving drugs from concept to the clinic. Curr Top Med Chem 2010; 10(1): 127-41.
[http://dx.doi.org/10.2174/156802610790232251] [PMID: 19929824]

[71]   Vijayakrishnan R. Structure-based drug design and modern medicine. J Postgrad Med 2009; 55(4): 301-4.
[http://dx.doi.org/10.4103/0022-3859.58943] [PMID: 20083886]

[72]   Coussa RG, Kapusta MA. Treatment of cystic cavities in X-linked juvenile retinoschisis: The first sequential cross-over treatment regimen with dorzolamide. Am J Ophthalmol Case Rep 2017; 8: 1-3.
[http://dx.doi.org/10.1016/j.ajoc.2017.07.008] [PMID: 29260104]

[73]   Van Drie JH. Computer-aided drug design: the next 20 years. J Comput Aided Mol Des 2007; 21(10-11): 591-601.
[http://dx.doi.org/10.1007/s10822-007-9142-y] [PMID: 17989929]

[74]   Lv Z, Chu Y, Wang Y. HIV protease inhibitors: a review of molecular selectivity and toxicity. HIV AIDS (Auckl) 2015; 7: 95-104.
[PMID: 25897264]

[75]   Hirsch R, Ternes T, Haberer K, Kratz KL. Occurrence of antibiotics in the aquatic environment. Sci Total Environ 1999; 225(1-2): 109-18.
[http://dx.doi.org/10.1016/S0048-9697(98)00337-4] [PMID: 10028708]

[76]   Kolpin DW, Furlong ET, Meyer MT, *et al.* Pharmaceuticals, hormones, and other organic wastewater contaminants in U.S. streams, 1999-2000: a national reconnaissance. Environ Sci Technol 2002; 36(6): 1202-11.
[http://dx.doi.org/10.1021/es011055j] [PMID: 11944670]

[77]   Basavaraj S, Betageri GV. Can formulation and drug delivery reduce attrition during drug discovery and development-review of feasibility, benefits and challenges. Acta Pharm Sin B 2014; 4(1): 3-17.
[http://dx.doi.org/10.1016/j.apsb.2013.12.003] [PMID: 26579359]

[78]   Šunjić V, Parnham MJ. Organic Synthesis in Drug Discovery and Development InSignposts to Chiral Drugs. Basel: Springer 2011; pp. 1-12.

[79]   Gillespie P. Am Drug Disc 2007; 2: 14-7.

CHAPTER 3

Environmental Xenoestrogens: Developmental Effect On Changing Environment, Molecular Mechanisms, and Human Health

Atul Gupta[1] and **Imran Ahmad**[2,*]

[1] *CSIR-Central Institute of Medicinal and Aromatic Plants, Lucknow, India*

[2] *Isabella Thoburn College, Lucknow, India*

Abstract: Estrogens, including estrone, estradiol, and estriol, are the female sex hormones conscientious for the regulation and play a significant role in the developmental process of the feminine reproductive organs. It is used for hypogonadal, postmenopausal, and hormone replacement therapy, as drugs in oral contraceptives and the cure of hormone-dependent cancers, such as breast cancer, ovarian cancer, and prostate cancer, and many other hormone-based complications such as osteoporosis. Environmental xenoestrogens may be classified into two categories- natural (derived from plants or fungi) and synthetic, which include steroidal estrogens, pesticides, and industrial waste. Phytoestrogens are thought to be beneficial for humans, but many environmental pollutants, including pesticides, plastics, and chemicals, which can mimic estrogen compounds, may act like estrogen or could interfere in the mechanism of action of natural estrogens and thus disturb the endocrine processes; such substances are called endocrine disruptors. In the last decade, concentrations of synthetic estrogens have increased rapidly in soil and water worldwide; synthetic xenoestrogens have attracted significant attention. In this chapter, the severe effects of xenoestrogens on human health have been highlighted.

Keywords: Breast cancer, Carcinogenesis, Early adolescence, Endocrine disruptors, Endometrial cancer, Environmental pollutants, Menopause, Mycoestrogens, Osteoporosis, Phytoestrogens, Steroid hormones, Xenoestrogens.

INTRODUCTION

Estrogens

The term 'estrogen' in American English or 'oestrogen' in British English is named after its importance in the oestrous cycle. Estrogens are steroidal hormones

---

* **Corresponding author Imran Ahmad:** Isabella Thoburn College, Lucknow, India; E-mail: imranlcc@gmail.com

Tahmeena Khan, Abdul Rahman Khan, Saman Raza, Iqbal Azad and Alfred J. Lawrence (Eds.)

or similar compounds which are responsible for the development of women's reproductive system and other sexual characters. They are responsible for a wide range of physiological functions. Along with progesterone, the estrogen hormones control menstruation. They play a very significant role in preventing infection and inflammation, reducing mental stress, strengthening bones, controlling cholesterol, and increasing sex drive. They also increase the glow of skin and hair, because estrogen increases the level of collagen of the skin, which reduces wrinkles and makes the woman look attractive and helps in blood clotting, and maintains the smoothness of the vagina [1, 2]. The term estrogen can also be used for any compound that is either natural in origin or synthetic but exhibits a similar effect as natural hormones. Estrogens, such as estrone (1; E1), estradiol (2; E2), estriol (3; E3), and estertrol (4; E4), which is another type of estrogen produced only during pregnancy, are steroidal core molecules (Fig. **1**). The steroid core represents a structure that contains a C18 nucleus, consisting of four fused rings out of them three are six-membered (rings A, B, and C) and one is five-membered (ring:D) [3, 4]. Estrogen is usually formed in women in the placenta and ovary. The most important form of estrogen is the 17β-estradiol with estrogenic hormonal activity, produced and secreted by the ovaries of conceiving women, which plays an important role in the development of the baby and then protecting it from miscarriage. Estrone and estriol are primarily formed in the liver from estradiol [5].

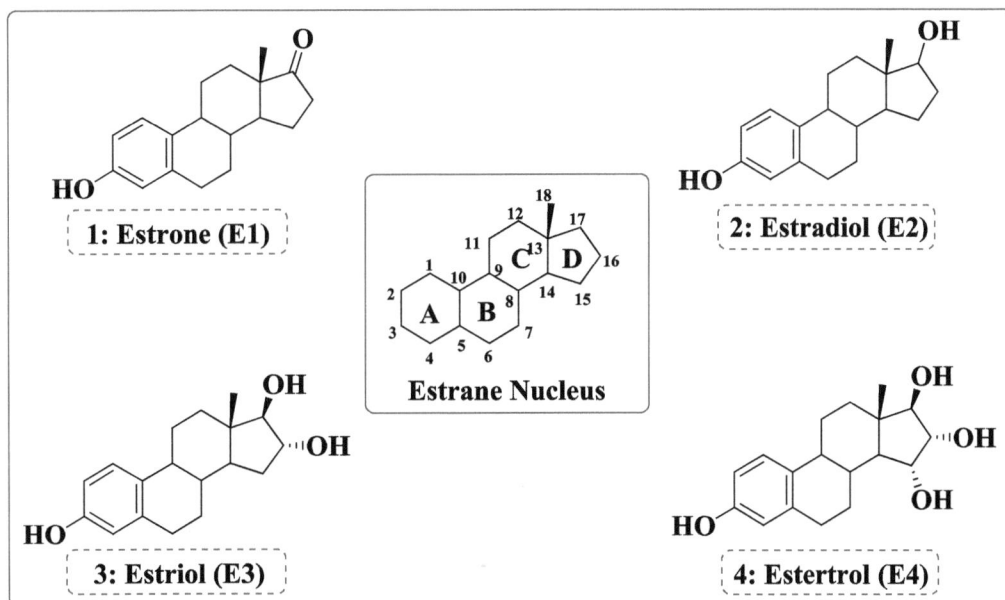

Fig. (1). Molecular structures of estrone (E1), estradiol (E2), estriol (E3), and estertrol (E4). The estrane nucleus is shown for reference.

Biosynthesis of Estrogens

The syntheses of sex hormones are regulated by the Gonadotropin-releasing hormone. The Gonadotropin-releasing hormone when released regulates the synthesis of sex hormones. The production of steroid hormones begins with cholesterol. When in the pituitary gland, GnRH induces the release of luteinizing hormone (LH) and follicle-stimulating hormone (FSH), after this process LH binds to its target cells and increases the expression of steroidogenic acute regulatory proteins (StAR). Biosynthesis involves cytochrome P50 and hydroxysteroid dehydrogenase (HSD) members. In corpus luteum, P450scc and 3beta-HSD, act as catalysts to reach progesterone from cholesterol, and they also catalyze P450arom (aromatase) in granulosa cells, from 19-carbon (C19) compounds to estrogens paves the way for biosynthesis, initiating the synthesis of theca cells at P450 [6, 7]. In the biological system, estrogen syntheses follow two routes:

Glandular Estrogen Synthesis

This type of synthesis occurs in granulosa as well as theca cells present in the ovaries. This type of synthesis also occurs in the corpus luteum. Luteinizing hormone stimulates granulosa cells to produce pregnenolone. Pregnenolone reaches the surrounding cells via these cells. After which Theca cells express 17, 20-lyase and 3-beta-hydroxisteroid dehydrogenase (3β -HSD), then metamorphose into androstenedione contraception via dehydroepiandrosterone.

Very often most of the androstenedione returns to granulosa cells, after which it is converted to estrone by the aromatase enzyme. After this process, the estrone is converted into 17β -estradiol with the help of a 17β -hydroxysteroid dehydrogenase enzyme. The expression of aromatase and 17β -hydroxysteroid dehydrogenase requires stimulation of follicle-stimulating hormone) (Fig. 2).

Extra-glandular Synthesis

It occurs with the expression of aromatase in non-gonadal sites and facilitates peripheral aromatization of androgens to estrone. The detailed chemistry of biosynthesis of estrogens is depicted in Fig. (2).

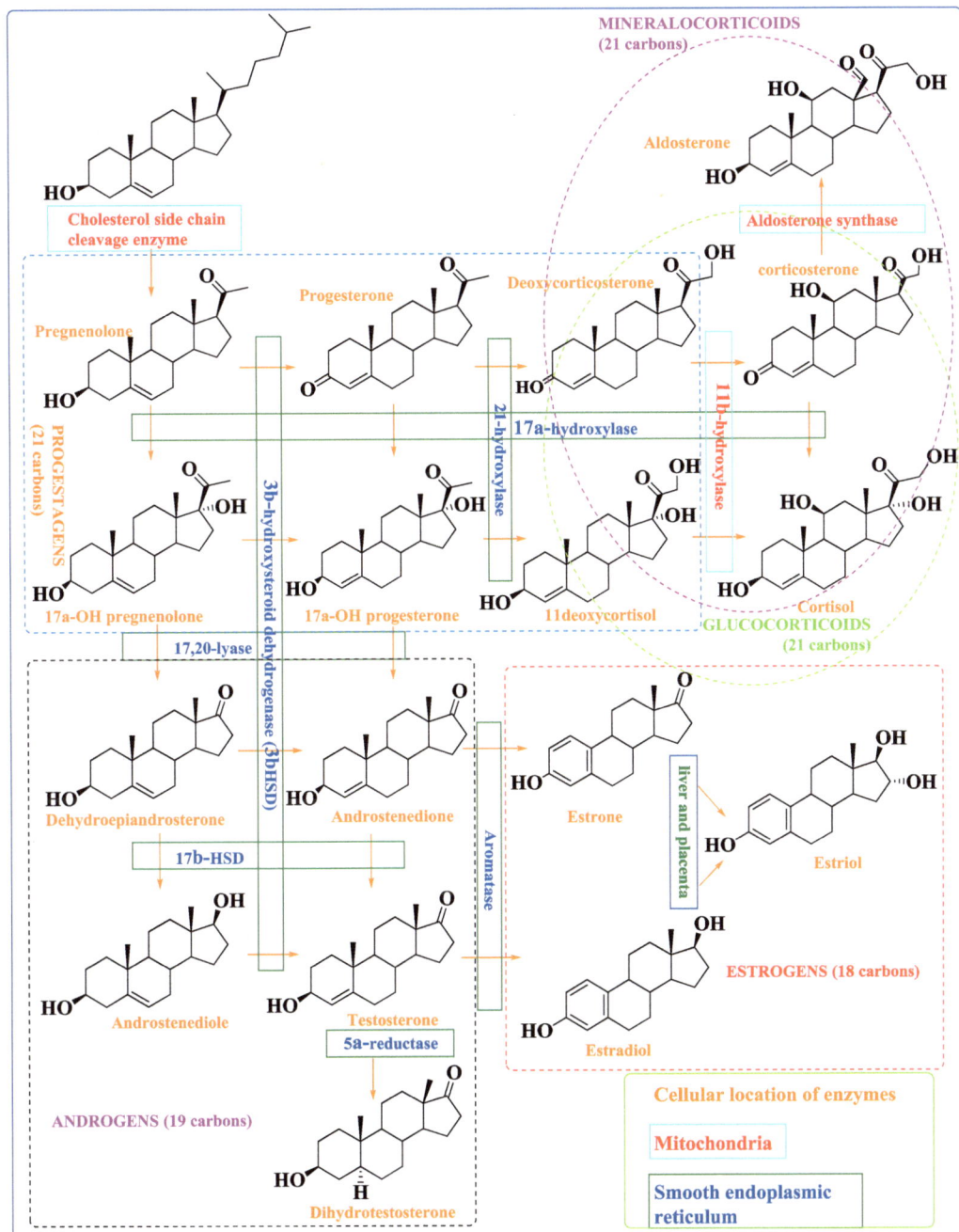

Fig. (2). Steroidogenesis [8], showing biosynthesis of estrogens at the bottom of the right.

Classification of Estrogens

It occurs with the expression of aromatase that facilitates peripheral aromatization of androgens to estrone in non-gonadal sites. The detailed chemistry of the biosynthesis of estrogens is depicted in Fig. (**2**) showing biosynthesis of estrogens at the right bottom.

Conventionally, all estrogens can be classified based on their origin. These can be divided into two major groups, *viz.* synthetic and natural estrogen. The synthetic estrogens are further divided into steroidal and nonsteroidal estrogens, based on their basic skeleton and atomic composition. Natural estrogens can also be divided on a similar basis but are mainly divided into two groups, *viz.* mammalian and phytoestrogen, based on their source. Fig. **3** is the diagrammatic representation of estrogen classification. In addition to the synthetic and natural estrogens, there are other classes of estrogens, known as xenoestrogen and mycoestrogen, which we have not covered here.

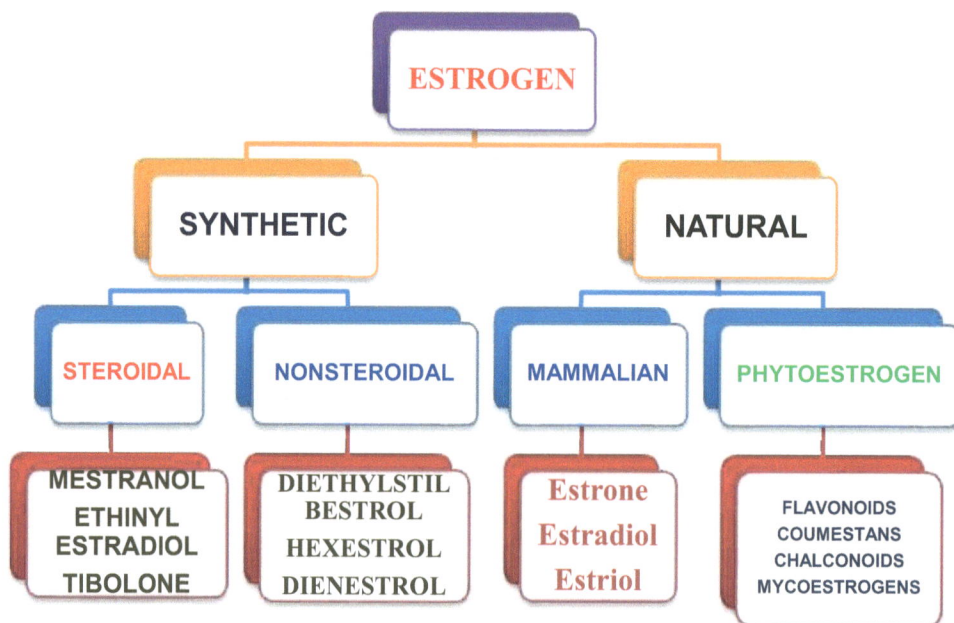

Fig. (3). Classification of estrogens.

Steroidal Estrogens

The naturally occurring endogenous estrogens, such as estrone (1), estradiol (2), and estriol (3), are found in humans and other mammals. Estradiol (1) is the predominant estrogen responsible for the estrogenic activity and in the determination of absolute serum levels.

Estradiol has poor bioavailability due to rapid metabolism. The addition of a 17α-alkyl group increases its bioavailability. Most of the therapeutically useful steroidal estrogens are produced semi- synthetically from natural precursors, such as diosgenin (5), a plant sterol [9, 10]. Estrone (1), estradiol (2), and estriol (3) are medically approved drugs.

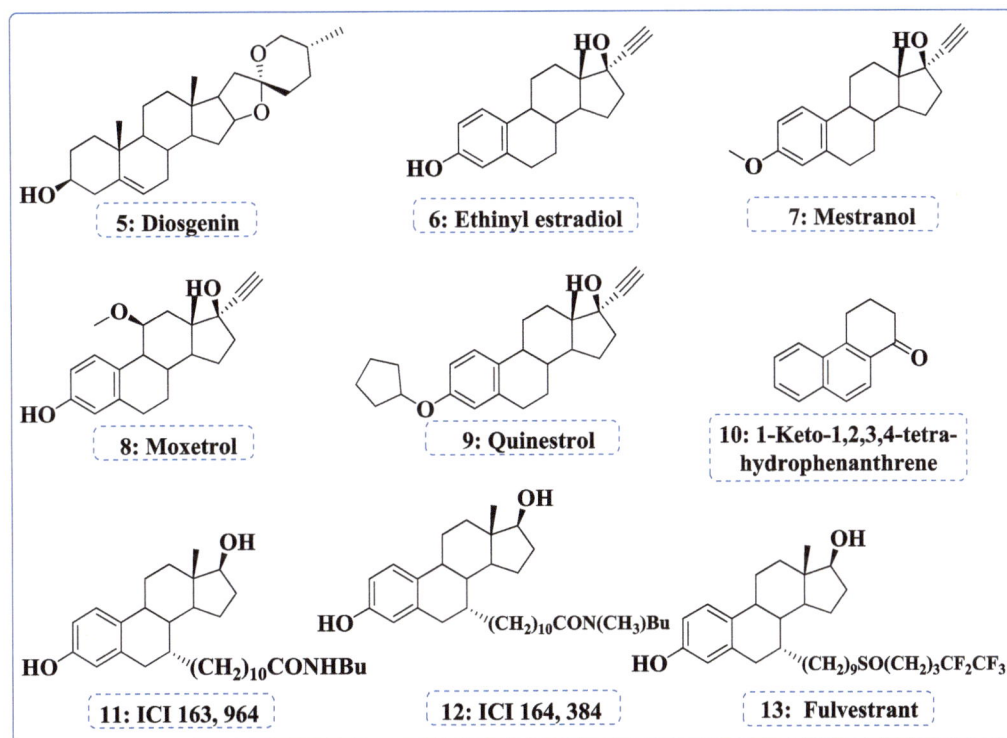

5: Diosgenin   **6: Ethinyl estradiol**   **7: Mestranol**

8: Moxetrol   **9: Quinestrol**   **10: 1-Keto-1,2,3,4-tetra-hydrophenanthrene**

11: ICI 163, 964   **12: ICI 164, 384**   **13: Fulvestrant**

Fig. (4). Some synthetic estrogens/antiestrogens.

Different long-chain carbon esters of estradiol (2) have been synthesized, *viz.*, estradiol cypionate, estradiol benzoate, estradiol acetate, estradiol undecylate, polyestradiol phosphate and, estradiol valerate, which behave as prodrugs to estradiol, have greater half-life than estradiol, and are medically important molecules [11]. Ethinyl estradiol (6; EE) is one of the more potent synthetic analogues of estradiol derivative that has been used for estrogen replacement therapy (ERT). Mestranol (7), moxestrol (8), and quinestrol (9) are the derivatives of EE and used clinically [12]. Furthermore, various 7-alkylamide derivatives of 17β-estradiol were synthesized; ICI 163,964 (11), ICI 164,384 (12), potent antiestrogenic compounds [13, 14] and sulfoxide derivative, Fulvestant (13; Facalodex), is being used to maintain estrogen levels in postmenopausal women and to treat estrogen receptor-positive metastatic breast cancer (Fig. **4**) [15 - 17].

Nonsteroidal Estrogens

The advancement in understanding the physicochemical properties of steroidal estrogens, such as 17β-estradiol, led to the discovery of the first synthetic estrogen, which was 1-keto- 1,2,3,4- tetrahydrophenanthrene (10, Fig. **4**) [17]; however, it had a weak binding affinity and possessed neither estrogenic nor antiestrogenic activity. Since then, many more nonsteroidal estrogens have been reported and are being explored extensively [18]. The molecules which do not possess a steroid core nucleus but structurally simulate the endogenous hormone 17β-estradiol (2), and exerting the similar effect of conventional estrogen, are called nonsteroidal estrogens. They are derived either from natural, semi-synthetic or synthetic sources. Compounds that act like natural estrogens, if they are obtained from a synthetic source, are called xenoestrogen, those derived from plants are known as phytoestrogen and those obtained from fungi are called mycoestrogen [19]. The synthetic nonsteroidal compounds which can mimic the action of natural estrogen estradiol (2), such as diethylstilbestrol (14; DES), hexestrol (15), dienestrol (16), benzesterol (17) and other related compounds, have been synthesized for their estrogenic activity (Fig. **5**) [20, 21].

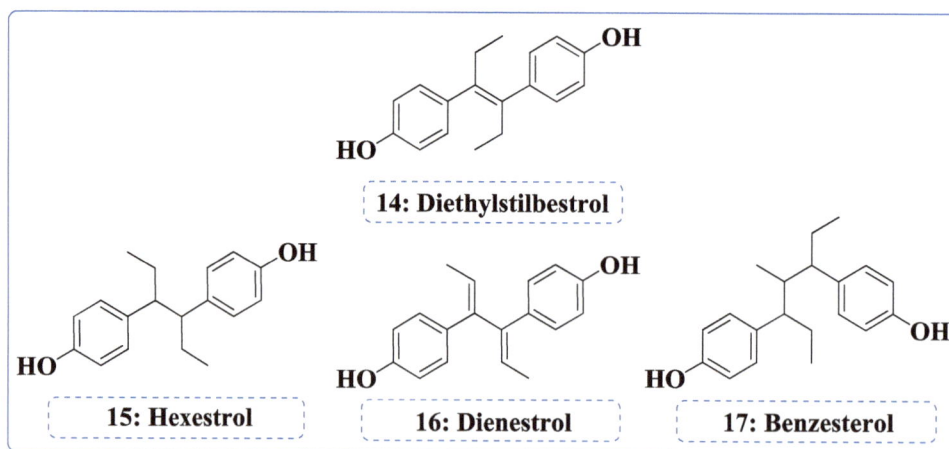

Fig. (5). Some nonsteroidal synthetic estrogens.

Several other naturally occurring chemical compounds that are structurally similar to estradiol (2) and DES (14), showing estrogenic activity, without possessing a steroidal framework, are examples of nonsteroidal estrogens. They have different core structures, such as flavonoids, isoflavonoids, coumestans and lignans. Phtoestrogens are phenolic compounds or their derivatives, such as coumestans, flavonoids and isoflavones, which are non-steroidal. The most explored phytoestrogens are isoflavones, commonly found in soy plants and red clover. The ability of these plant-derived compounds to cause estrus in animals was observed in the mid-1920s. This led to the development and modification of nonsteroidal compounds like estradiol for better estrogenic activity.

Among the nonsteroidal compounds obtained from plants, formononetin (18), isoformononetin (19), daidzein (20), genistein (21), naringenin (22), glabridin (23), resveratrol (24), medicarpin (25) and glycinol (26) are notable examples. Equol (27), the active metabolite of daidzein, is a phytoestrogen, and structurally similar to 17 β-estradiol (2) and showed significant estrogenic activity (Fig. 6) [1, 22 - 28].

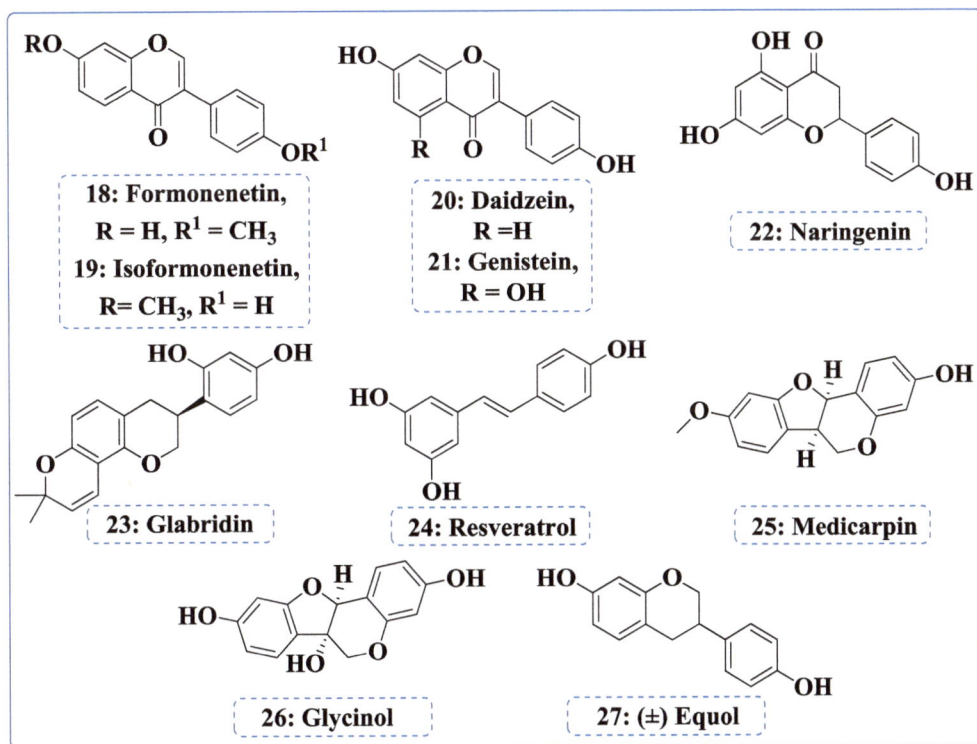

Fig. (6). Some naturally occurring phytoestrogens and their active metabolite.

Functions and Mechanism of Action of Estrogens

Estrogens are hormones that are primarily required for the development, growth, and maintenance of different tissues in both females and males, especially in the development of sexual characteristics in women. Since they play an important role in many physiological processes hence, they find use as signalling target molecules for different pathways. Thus, they have a great impact in the medical field for the treatment and management of hormone-dependent disorders, as anti-osteoporotic agents, contraceptives, and in the treatment of various cancers (Fig. 7) [29]. Estrogens act as signalling molecules and through the bloodstream they reach a variety of target tissues where estrogen receptors are abundant, such as in the breast and uterine cells, where they bind to the receptor and exert their effects [30 - 36].

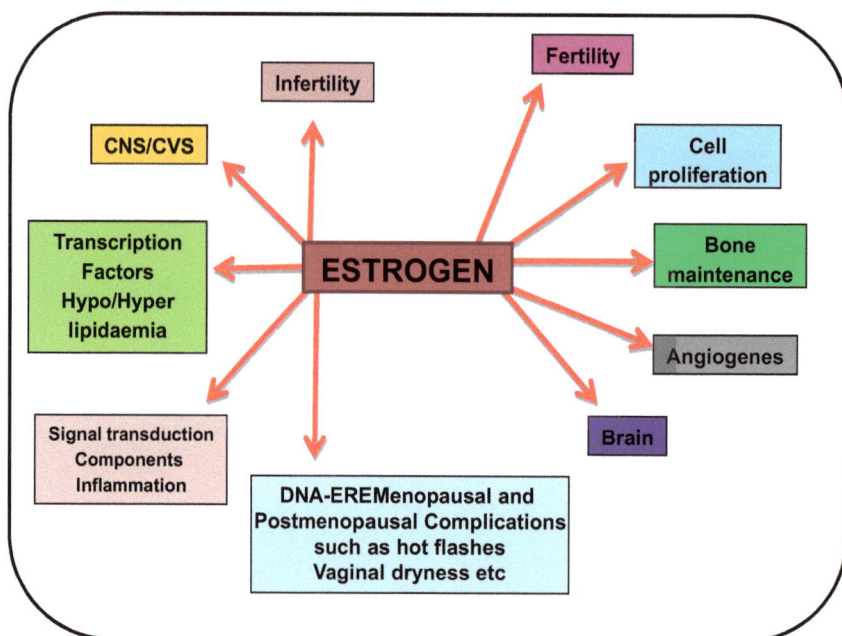

Fig. (7). Biological activities of estrogen.

It was a path-breaking discovery in this field of estrogen action when Elwood Jensen (97) discovered an estrogen binding protein in the late 1950s which was later recognized as estrogen receptor alpha (ERα) [37]. In the early 1990s, another form of estrogen receptor, *i.e.*, ERβ was discovered by knocking ERα from a mouse model. Today, these two receptors, belonging to the nuclear receptor superfamily, are known targets for estrogen action [38, 39].

Estrogen Receptors

The estrogen receptors are a group of proteins present in the cells, where they bind to estrogens. For the cellular signalling of estrogens, two estrogen receptors (ERs) are responsible one is ERα (NR3A1) and the second one is ERβ (NR3A2). Both the isoforms of ERs belong to the nuclear receptor (NR) family of transcription factors. Like many other members of the NR family, ERs have unique domains that are structurally and functionally conserved (Fig. **8**) [40 - 42].

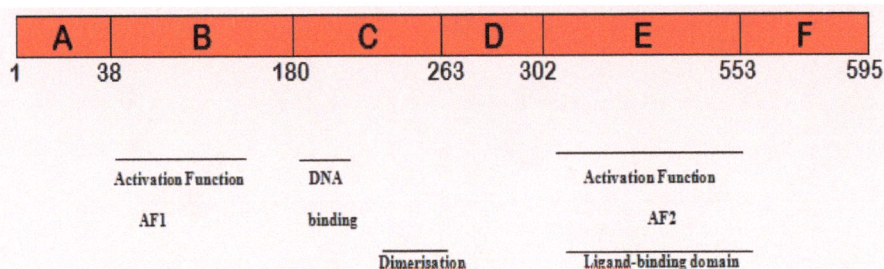

Fig. (8). Structure of estrogen receptor (ER); A to F represent different domains of the ER. Numbers represent amino acids from amino to carboxy-terminus.

When estrogens bind to ERs and activate them, then only they translate into the nucleus and subsequently bind to DNA in the nucleus and regulate the activity of DNA genes. Along with this, they are also responsible for many other functions independent of DNA binding [43].

The estrogen receptors (ERs) consist of 3 functional and structural domains. Out of these three domains two domains are amino (N-terminal) modulatory and the third one is the DNA binding domain, having specific binding pockets for target DNA. The amino-terminal domain is large and contains 180 amino acid residues. The DBD is small and present at the C-terminus which consists of approximately 65 residues; occupying the central portion of the protein [44].

Estrogen receptors use two distinct domains for transcriptional activation which enable them to stimulate transcription. In this process, TAF-1 is present in the amino-terminal (N-terminal) domain and TAF-2 in the estrogen-binding domain. For the activity of TAF-2, a region of the hormone-binding domain at the carboxy-terminus (C-terminus) is required (as between residues 538–552 in the mouse estrogen receptor). This process is conserved in many nuclear hormone receptors.

For the hormone-receptor complex formation, it is imperative to recognize DNA sequences upstream of transcriptional onset sites. The separate domains are identified in the ERα and ERβ as shown in Fig. (**9**). The important difference between both isoforms is in their LBP with the substitution of Leu 338 in ERα with Met 384 in ERβ Fig. (**9**) [29, 39].

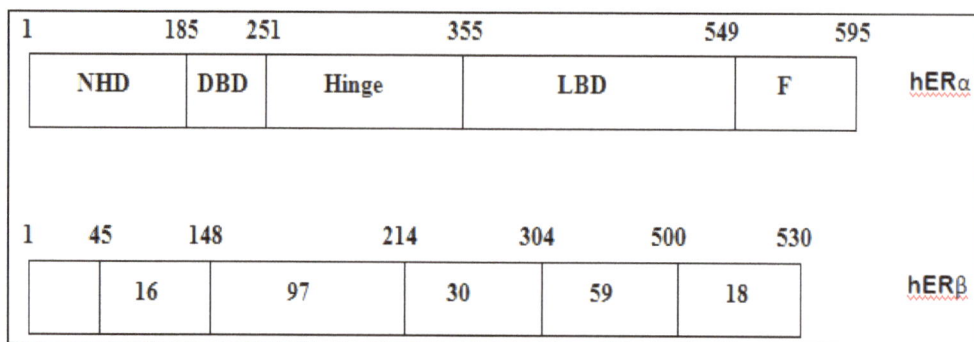

1	185	251	355	549	595	
NHD	DBD	Hinge	LBD	F		hERα

1	45	148	214	304	500	530	
	16	97	30	59	18		hERβ

Fig. (9). Comparison of amino acids sequence of the ERα and ERβ receptors.

Estrogens affect gene expression as they bind and activate their receptors, located primarily in the nucleus. The binding of ligand to the receptor leads to dimerization of the receptor. It then facilitates the binding of the receptor dimer directly to the promoter region of DNA either by activating or suppressing the process of transcription. The two isoforms of ER are very specific in their functions and act as transcription factors after binding their specifically designed ligands [45 - 49]. It is believed that ER- ligands enter the cell either by passive diffusion or through the cell membrane where they bind to estrogen receptors which are present inside the cytoplasm. After binding with receptor proteins, they undergo a series of conformational changes, move into the nucleus, and get dimerised.

Inside the nucleus, estrogen receptors either attach directly to DNA through specific ligands or in combination with other transcription factors, such as activator protein 1 (AP1) and specific protein 1 (SP1). To effectively conduct gene expression, the ER takes the help of other co-regulatory protein complexes, then signalling the gene transcription machinery. The genomic and non-genomic mechanisms of estrogen action are depicted in Fig. **10**.

In the genomic pathway, the nuclear receptor (NR) is captured by cognitive hormones leading to the regulation of genes either upstream or downstream. In the non-genomic pathway, steroidal hormones may bind to a cystic membrane receptor (which in some cases may or may not be a membrane-associated nuclear receptor), which can cause the onset of a rapid response that is either by giving signal directly to second-messenger leading to biological response (s) or through intonation of genomic responses Fig. (**10**) [50, 51].

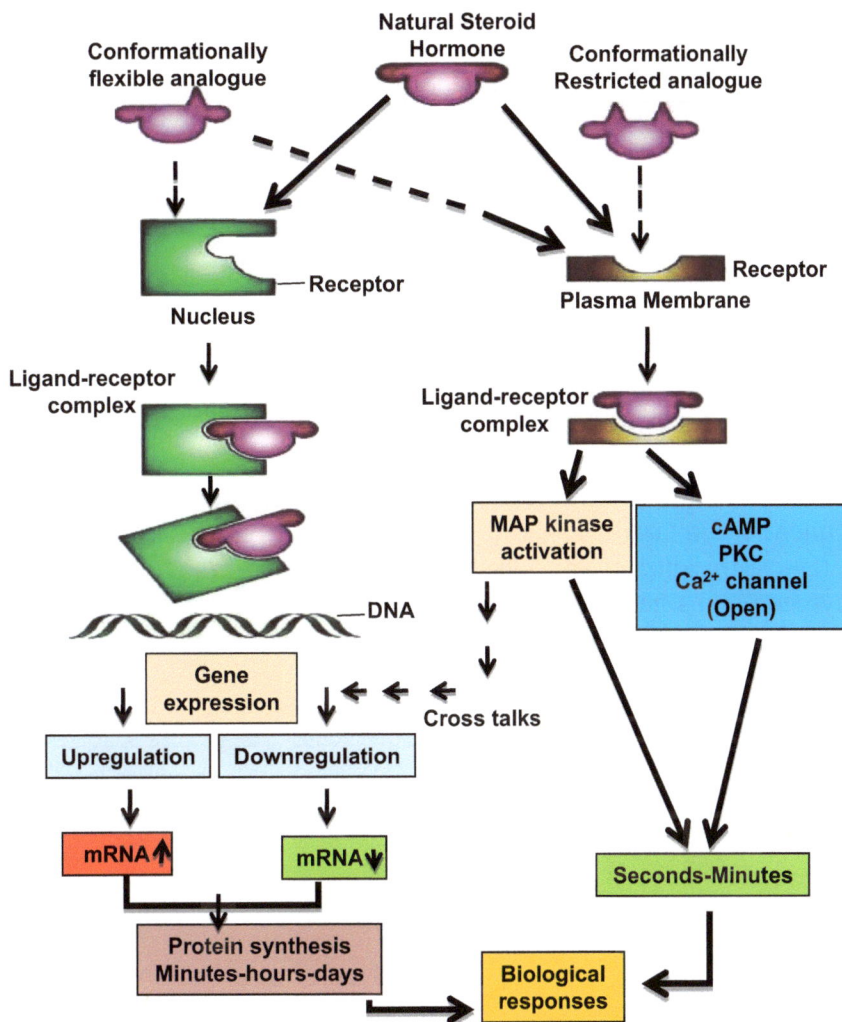

Fig. (10). Genomic and nongenomic pathways of estrogen receptor ligands.

Xenoestrogens

The literal meaning of 'xeno' is foreign, therefore xenoestrogen means foreign estrogens. They structurally simulate natural estrogens and hence can bind with estrogen receptor sites to produce potentially hazardous effects. These chemicals adversely affect the endocrine system, which is why they are called endocrine disruptors/inhibitors (EDs). They use various mechanisms to exert negative effects: either by linking to the related hormone receptors or by disrupting the cell signalling pathway directly; or by having a direct effect on the central nervous system (CNS) and neuroendocrine systems. Apart from these mechanisms they may also show negative effects by down-regulation of hormone synthesis, or through their fatal effects on malignant tissues.

Xenoestrogens may be categorized into two types:

- Natural
- Synthetic

Natural Environmental Estrogen

The best-known chemicals belonging to this group are phytoestrogens (Table **1**) and mycoestrogens (produced by fungi), which show weaker activity as compared to natural steroidal hormones. Phytoestrogens are taken through dietary sources and frequently consumed in daily life. In higher concentration phytoestrogens exhibit their estrogenic effect, whereas antiestrogenic in low concentration [52 - 55]. Phytoestrogens have shown advantageous effects against cancer, cardiovascular disorder, and osteoporosis [56 - 58].

Table 1. Some important phytoestrogens which have an estrogenic effect.

Phytoestrogens	Daidzein, genistein, formononetin, biochanin-A, prunetin, pratensein, glycitein, equol, desmetilangolestin, enterolactone, enterodiol, matairesinol, zearalenone

Synthetic Xenoestrogens

Many synthetic xenoestrogens, such as pesticides (DDT, methoxychlor, organohalogens), industrial by-products (polychlorinated biphenyls, alkylphenols, bisphenol A), and drugs that show similarities in their functions to endogenous estrogen estradiol can persist for a long time in the environment. These xenoestrogens accumulate in the food chain and human biological matrices [59].

Most of these EDCs often accumulate in fat tissue and may persist in the body for long periods and cause destructive effects [54, 56, 60 - 65]. In humans, they have been found to disrupt ovarian and testicular histopathology, gonadal intersex, reduction in gonad size, vitellogenin induction, and alter sex ratios [66 - 68]. Various environmental pollutants act as endocrine disrupters, which can interfere in the mechanism of natural estrogen, leading to destructive effects on human health (Tables **2**, **3** and Fig. **11**) [69 - 71].

Table 2. Some synthetic xenoestrogens which have an estrogenic effect.

Type	Examples
Pesticides	Dichlorodiphenyltrichloroethane (DDT), methoxychlor, endosulfan, 2,4-dichlorophenoxyacetic acid, alachlor, aldicarb, amitrol, atrazine, benomyl, dibromo chloropropane, carbaryl, chlordane, ethyl parathion, heptachlor, kepone, ketoconazole, lindane, methomyl, permethrin, malathion, trifluralin, vinclozolin
Phthalates	Diethylhexyl phthalate, butyl benzyl phthalate, di-n butyl phthalate, di-hexyl phthalate, di-propyl phthalate, dichloro hexyl phthalate, diethyl phthalate
Industrial products	Bisphenol A, polybrominated biphenyls
Organohalogens	Dioxins, furans, polychlorinated biphenyls, hexachlorobenzene, pentachlorophenol
Heavy metals	Arsenic, cadmium, uranium, lead, mercury
Drugs	Oral contraceptives, diethylstilbestrol, cimetidine

Table 3. List of xenoestrogens which may commonly be used in day-to-day life and act as endocrine disruptors (EDs).

Chemical	Use(s)
Alkylphenols	Intermediate chemicals used in the manufacture of other chemicals
Atrazine [28]	Commonly used as a weed killer
Butylated hydroxyanisole (BHA) [29]	Used as a food preservative
Bisphenol A [30]	Commonly used monomer for polycarbonate plastic and epoxy resin
Dichlorodiphenyldichloroethylene [31]	One of the breakdown products of DDT
Dieldrin [32]	A banned insecticide in many countries
DDT [33]	A banned insecticide
Endosulfan [34]	Widely banned insecticide, not yet banned in India
Heptachlor [35]	Restricted insecticide, not in India
Lindane [36]	A widely restricted insecticide
4-methylbenzydene camphor [37]	Widely used in sunscreen lotions
Pentachlorophenol [38]	Used as general biocide and wood preservative
Polychlorinated biphenyls	Formerly used in electrical oils, lubricants, adhesives, and paints banned these days in many countries
Parabens	Commonly used in lotions
Phthalates	Used as plasticizers
Phenosulfothiazine	Used as a red dye
Propyl gallate	Used as an antioxidant to protect oils and fats from oxidation.

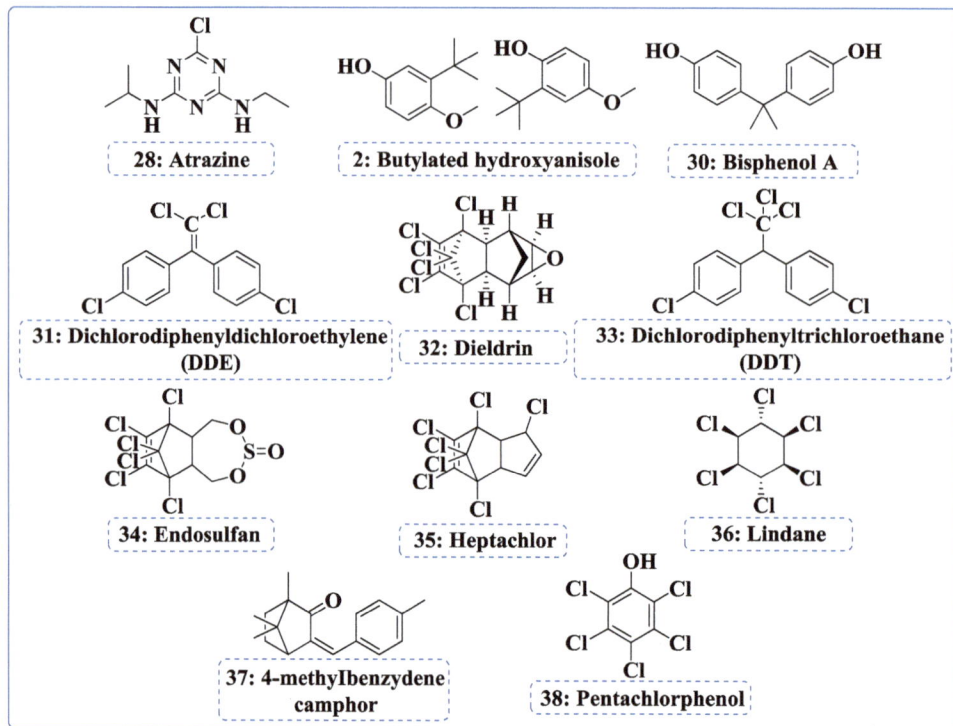

Fig. (11). Some synthetic endocrine disruptors (EDs).

Mechanism of Action of Xenoestrogens

a. Binding Directly to Estrogen Receptor

Since they are structurally similar to natural endogenous estrogens, they act like natural estrogens. Most xenoestrogens bind to estrogen-receptors, and they may or may not compete with estradiol for binding. Most xenoestrogens bind to the estrogen-receptors and may or may not compete with estradiol for binding. In the same way that E2 functions, these compounds also promote active ER conformations and force them to promote their functions, resulting in the regulation of target genes [70, 72, 73] (Fig. **10**) [50, 51].

b. Other Plausible Mechanisms

Some endocrine-disrupting chemical compounds may impair the connection of natural hormones to plasma transport proteins *viz.* displacement of thyroid hormones by dioxin and organochlorine results in decreased hormone levels [74].

Decomposition products of DDT, methoxychlor and some other xenoestrogens may promote the activity of estrogen receptor (ER). They are ER agonists or may suppress the activity of androgen receptor (AR), *i.e.*, AR antagonists [75, 76]. Studies have revealed that the *in-vivo* decomposition product of DDT is DDE which has greater affinity with AR because it can bind to this receptor even at low concentration, whereas at higher concentration it shows affinity with AR as well as with ER. Thus, one might hypothesize that endocrine disintegration occurs at a higher concentration resulting from a higher passage perturbation, whereas at lower concentrations, the probability becomes low since only one of the two receptors may be involved [77, 78].

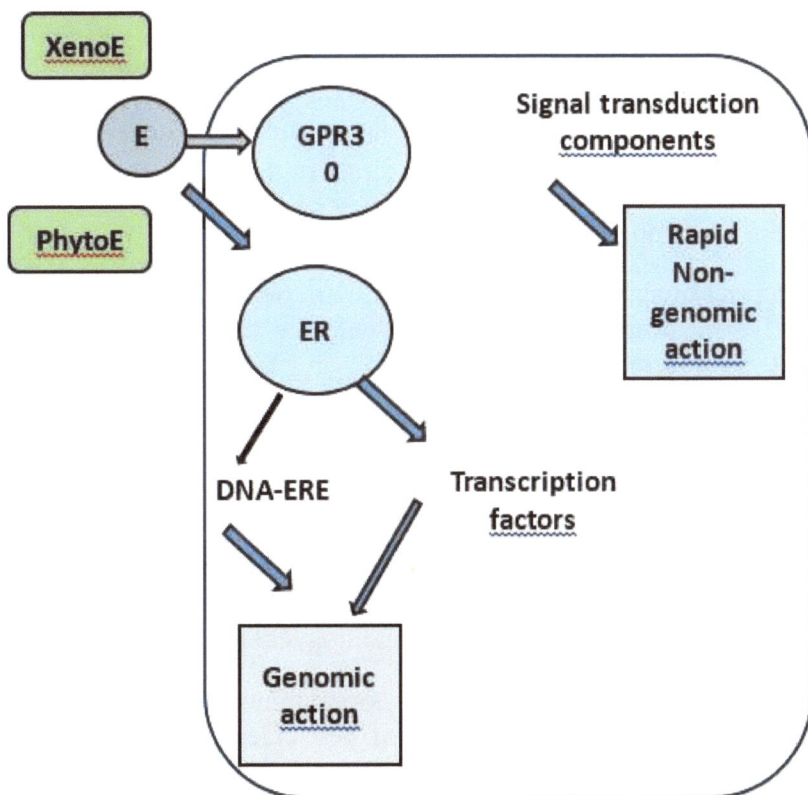

Fig. (12). Diagram showing the mechanism of action of estrogenic compounds present in our diet.

Sources of Xenoestrogens

There may be various sources that are responsible for the entry of xenoestrogens into the environment; including industrial, agricultural, and chemical products, such as foods, drugs, cosmetics, insecticides, herbicides, plastics, oils, paints, and adhesives.

Some major sources by which xenoestrogens enter the environment are discussed here.

Water

Water is an integral part of life; studies suggest that both surface and underground water is contaminated with xenoestrogens [79]. Since the water treatment plants currently used for wastewater treatment are not designed to eliminate hormonal pollutants, agricultural chemicals, and pharmaceutical runoff, therefore, the discharge of untreated waste from different sources is the possible way by which water is contaminated. Population near the sites where pesticides are manufactured /used/disposed of are more exposed to xenoestrogens. Studies reveal that drinking water, usually stored in cans, plastic bottles, buckets, and drums, is not safe because phthalates may leach out in the water on prolonged storage, especially at high temperature or in direct sunlight.

Chemical Pesticides and Fertilizers

Fertilizers are used in agricultural land to maintain or improve the quality of the soil, while different pesticides are used to enhance the production process of plants, as well as improve plant growth.

One of the popular xenoestrogens is DDT, which was banned in the United States. Unfortunately, it is still used agriculturally for some products in many countries, as it is a very stable compound, has a long life, and is therefore present in the land for years; hence its ban is of no use because, despite the ban, DDT is still regularly harming people worldwide. Not only DDT but many other agricultural chemicals contain xenoestrogens; among the common products known to contain xenoestrogens are atrazine, endosulfan, and methoxychlor.

Several pesticides, such as organochlorides, organophosphates, carbamates, etc. also contain xenoestrogens. The presence of pesticides, herbicides, fungicides, and fertilizers in water is due to the erosion of rain and runoff from landfills and agricultural land. They are also present in the food supply - in plants, animals, fish, and grains. They can persist in soil and water and may bio-accumulate in organs, where they are stored in fatty tissue such as breasts. They can cause serious effects on human health [80].

Non-Organic Food Products

To enhance the appearance of vegetables, fruits, cereals, etc., commercial food producers spray pesticides. It is clear from various scientific reports that approximately only 1-2% of sprayed pesticides fall on pests, whereas crops are more exposed. These pesticides get deposited on the crop and may enter human and animal bodies.

Similarly, non-organic livestock is typically given hormones to help them grow large and fat. The same problem that impacts cattle meat also applies to the milk that they produce. A similar problem is seen with other dairy products as well as poultry. Thus, non-organic food may be a major source of xenoestrogens [81].

Plastics

One of the major sources of xenoestrogens is plastics, especially soft plastic, and its by-products. Plastic is commonly used in the form of cups, plates, airtight food storage containers, etc. But plastic is a petroleum by-product that leads to leakage of xenoestrogens in food. Plastics contain the following chemicals which may leach out at elevated temperature or in response to other stimuli and act as xenoestrogens [82]:

- **Bisphenol A** (monomer for polycarbonate plastic and epoxy resin; antioxidant in plasticizers)
- **Phthalates** (plasticizers)
- **DEHP** (plasticizer for PVC)
- **Polybrominated biphenyl ethers (PBDEs)** (flame retardants used in plastics, foams, building materials, electronics, furnishings, motor vehicles).
- **Polychlorinated biphenyls (PCBs)**

Cosmetics and Toiletries

Xenoestrogens negatively impact human health, and their artificial sources are also potentially harmful to humans. These endocrine-disrupting chemicals are most easily absorbed due to the abundant use of skincare products and enter the bloodstream [83]. Nowadays, there are a lot of beauty products in the markets. These beauty products use various chemicals, usually above the prescribed limits. The ingredients used in beauty products contain phthalates, parabens, triclosan, etc. which act as xenoestrogens. These cosmetic products contain xenoestrogens which can enter the bloodstream through absorption. They completely bypass the liver and thus have a very rare chance of getting eliminated from the body. The largest organ in the human body is the skin and thus it is most exposed to cosmetics. So, what we apply to our skin is even more important for our health than what we eat.

Many reports showed that approximately 60% of the chemicals (Xenoestrogens) enter the bloodstream through skin absorption; the rate of absorption through the skin on our face is five to ten times higher than elsewhere on our bodies. This means that every time we use cosmetics like face toner, under-eye concealer, foundation, nail polish, perfumes, soaps, shampoos, etc. we allow our skin to absorb these endocrine-disrupting chemicals.

The following ingredients are commonly used in cosmetics and other beauty products:

Phthalates

Phthalates are plasticizers that are used in various products to enhance skin quality. They are used in skincare to moisturize and soften the skin.

Parabens

Parabens are used to increase the quality of personal care products. Due to their antimicrobial and preservative tendencies, they are added to various products such as shampoos, conditioners, and scrubs, etc. Just like phthalates, they are also xenoestrogens. Parabens are easily absorbed into the bloodstream through the skin. Therefore, their use in skincare products gives them an easy route to our bloodstream.

Triclosan

Triclosan was originally developed for use in surgical scrubs due to its strong antimicrobial tendency. But various researches have shown that triclosan affects the production of thyroid hormones in animals.

Household Cleaners and Kitchen Ware

Household cleaners have been one of the trademarks of a healthy home for a long time, but many household cleaners contain harmful chemicals which may act as xenoestrogens. Laundry products such as detergents, fabric softeners, dryer sheets, etc. contain harmful chemicals which may act as endocrine disruptors. The floor cleaner and other products which contain xenoestrogens are even more harmful than dirt. In these products, many have volatile organic compounds (VOCs) that can cause cancer and a variety of negative effects in which allergies, respiratory problems, hormone disruptions, and eczema are common. These chemicals are used extensively in various cleaning products. After their usage, these harmful chemicals remain in the air for a long time and cause harm.

Similarly, non-stick cookwares are coated with Teflon polymer which has been proven to cause harmful effects on human health. Teflon polymers can increase the risk of cancer, cause fertility problems, and may damage the liver [84].

Pharmaceutical Products

Advances in civilization, coupled with rising population levels, have resulted in increased use of birth control pills and spermicidal gels. These pharmaceutical products contain synthetic estrogens, which mimic natural hormones and act as xenoestrogens [85].

Adverse Effects of Xenoestrogens

The potential hazardous effects of xenoestrogens on human health and biological organisms are a global concern. They have been implicated in a variety of medical problems during the last two decades.

Xenoestrogens exert the following harmful effects on humans [71, 86, 87]:

- Xenoestrogens and their metabolites are identified as cancer-causing agents in different organs *viz.* breast, uterus, ovary, testicle, prostate, and kidney [67, 86, 88].
- Based on different trials on animal and human models, the role of xenoestrogens in precocious puberty has been confirmed. They are responsible for early puberty via their estrogenic or antiandrogenic effects and by increasing the production of gonadotropin-releasing hormones (GnRH) [68, 89 - 91].
- They may act as artificial messengers and disrupt the process of reproduction.
- Xenoestrogens may cause developmental neurotoxicity. Some research groups have addressed this issue prominently in their research; Epidemiological associations have prominently highlighted perinatal or pre-natal risks to PCBs, pesticides, and polychlorinated dibenzofurans due to cognitive and behavioural defects [88, 92].
- Xenoestrogens may have adverse effects on cardiovascular risk factors, such as thrombosis and inflammation [93].
- Through various researches, it has been known that mere exposure to low amounts of BPA can also cause many lifestyle disruptions such as increased

susceptibility to obesity and diabetes later in life [94 - 96].

- Various researches have shown that exposure to environmental xenoestrogens leads to disruption or imbalance of systemic hormonal regulation of the skeleton, including bone modelling and remodelling, local hormones, and cytokine or chemokine release. This may lead to osteoporosis [97].

- Bisphenol-A (BPA) poses a greater risk to male workers and causes male sexual dysfunction across all domains of sexual function [98].

- Dibutylphthalate (DBP) is an environmental xenoestrogen, used as a plasticizer; it has negative effects on the reproductive system, as demonstrated by both *in vivo* and *in vitro* analyses [99].

- Xenoestrogens that are present abundantly in the atmosphere, water, food, air, cosmetics, etc., have harmful effects on human health. They are hazardous for cervical cancer development [100].

- Heavy metals affect human health, both through environmental as well as industrial exposure. There is evidence that metallo-xenoestrogens accumulate in several organs and are carcinogenic to humans. One of the important hazardous metals is Cd, which mimics the effect of natural estrogens in the uterus and mammary gland [101, 102].

- Nonylphenol and its derivatives that are used as emulsifiers, industrial surfactants, and in laboratory detergents, can alter the mechanism of natural estrogen and disturb the sexual developmental process of humans [103].

- In general, parabens are used as preservatives in various foods, cosmetics, and pharmaceutical products and are currently being used across the world. They easily penetrate the human body. They are endocrine-disrupting chemicals (EDCs), disrupting the fitness and functions of the body [104, 105].

Xenoestrogens: Implications for Risk Assessment

A lot of research has been done on the effects of consumption of compounds with estrogenic activity, on human health. Research reveals that xenoestrogens are likely to show both beneficial and adverse effects. Generally, compounds of 'natural' origin are considered beneficial, while synthetic compounds are considered harmful.

Although both phytoestrogens and xenoestrogens exhibit estrogenic activity as shown in *in vitro* and *in vivo* models, it has become a concept in society that phytoestrogens are generally useful, but it is a common practice to suspect the use of xenoestrogens. However, it may be arguable that the potential benefits of phytoestrogens are greater, and the adverse effects are minimal, while the opposite may be true for xenoestrogens.

Endocrine-Disrupting Chemicals affecting Reproduction

There are probably hundreds of chemicals responsible for endocrine-disrupting activity (Table **4**). Here we are discussing some examples of EDCs with *in vitro* models and/or endocrine- disruptive effects in humans within the scope of current human exposure.

Table 4. Xenoestrogens with endocrine-disrupting effects [106 - 119].

Compound(s)	Hormone system Affected	Mechanism if known
Benzenehexachloride (BHC)	Thyroid	-
1,2-dibromoethane	Reproductive	-
Chloroform	Reproductive	-
Dioxins and furans	Estrogen	Work as anti-estrogen through binding with Ah receptor, which then inhibits estrogen receptor binding to estrogen response elements, thereby inhibiting estrogen action
PCB, hydroxylated	Thyroid	Binds to thyroid hormone-binding protein, but not to the thyroid hormone receptor
Butylated hydroxyanisole (BHA)	Estrogen	Inhibits binding to the estrogen receptor.
Aldrin	Estrogen	Binds to estrogen receptors; competes with estradiol
Karate	Thyroid	A decrease of thyroid hormone in serum; direct effect on the Thyroid gland.
Ketoconazole	Effects on reproductive systems	-
Lindane (Hexachlorocyclohexane)	Estrogen/Androgen	Inhibits ligand-binding to androgen and estrogen receptors
Methoxychlor	Estrogen	Through mechanisms other than receptor antagonism. The precise mechanism is still unclear
Di-Ethylhexyl phthalate (DEHP)	Estrogen; Androgen	Inhibits binding to the estrogen receptor. Anti-androgenic
Benzophenone	Estrogen	Binds weakly to estrogen receptors, roles of its metabolite remain to be clarified
Bisphenol A (BPA)	Estrogen	Estrogenic; binds to estrogen Receptor
Benzo(a)pyrene	Androgen	Anti-androgenic
Nonylphenol, octylphenol	Estrogen	Estrogen receptor agonists; reduce estradiol binding to the estrogen receptor
Resorcinol	Thyroid	-

(Table 4) cont.....

Compound(s)	Hormone system Affected	Mechanism if known
Styrene dimers and trimers	Estrogen	Estrogen receptor agonists
Arsenic	Glucocorticoid	Selective inhibition of DNA transcription normally stimulated by the glucocorticoid-GR complex
Cadmium	Estrogenic	Activates estrogen receptor through an interaction with the hormone-binding domain of the receptor
Lead	Reproductive	-
Mercury	Reproductive/ Thyroid	-

Suggestions

Xenoestrogens enter the environment by industrial, agricultural, and chemical sources, and are used extensively in foods, medicines, cosmetics, pesticides, herbs, plastics, oils, paints, and adhesives, etc. Because we make extensive use of chemicals with estrogenic properties in our day-to-day life, it is not possible to avoid their risk. Nevertheless, we can reduce the side effects caused by following the following preventive measures:

- Use organic fertilizer as much as possible and minimize the use of synthetic pesticides, herbicides, and fungicides.
- Use natural pest control.
- While using foodstuff, vegetables, etc. wash them properly to get rid of pesticides.
- Do not switch to bottled water. Instead, install a reverse osmosis water system.
- Eat organic food as much as possible.
- Minimize the use of plastic containers in the microwave. Use glass or ceramics.
- Avoid plastic containers, especially to store drinking water in the sunlight.
- Avoid Teflon and other non-stick kitchen wares, instead, use cast iron.
- Avoid the use of household cleaners and use traditional household cleaners like baking soda, borax, and vinegar.
- Avoid packed food in tin cans.
- Avoid the use of fabric softeners and dryer as it puts petrochemicals right on the skin.
- Use organic soaps and toothpaste.
- Minimize the use of shampoos, creams, sunscreen lotions, and cosmetics that contain harmful parabens, as much as possible.
- Avoid nail polish and nail polish removers.
- Get rid of perfumes as they are petrochemical-based. Use only natural-based

perfumes. Increase the use of natural products as much as possible.
- Instead of medicines, use a simple method of birth control such as condoms.
- Try different detoxification methods for Xenoestrogens.
- As a strategy, the steps shown in Fig. (**12**) can be used to reduce human and animal contamination, thereby avoiding all types of malfunctions or diseases.

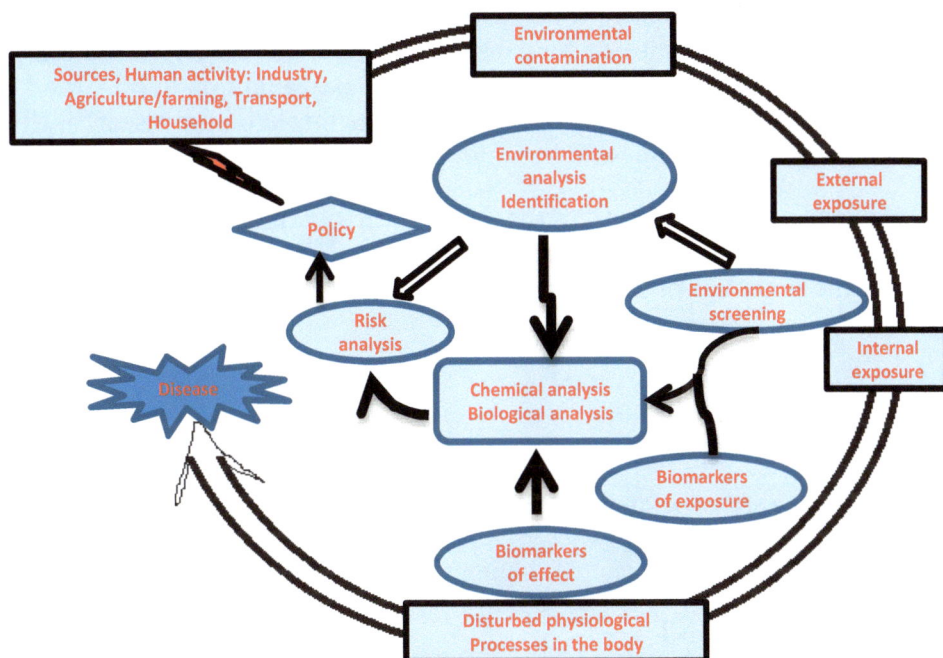

Fig. (13). Flow diagram depicting the strategy to be implemented for eliminating hormone-disrupting agents from the environment and living organisms.

CONCLUDING REMARKS

In conclusion, it is worthy to note that xenoestrogens are accountable for many disorders. It has been known from various research works that endocrine disruptors are harmful chemical substances that are harming the function of many endocrine systems through multiple mechanisms. Their role in early obesity and neurological and endocrine disorders in children is also increasing day by day which has become a matter of great concern to the world. Also, they contribute to complex diseases such as osteoporosis, estrogen-dependent cancers, diabetes, and cardiovascular disease, and hence they are a concern for the modern world. There should be collaborative efforts by environmental health experts and oncologists to combat the harmful effects of xenoestrogens. It is a harsh truth that on the one hand, food supplies energy to the body for normal function, while on the other hand, it is one of the most important sources of exposure to both natural and synthetic chemicals. Nowadays food contains harmful chemicals, causing many adverse effects on the health of humans, with which endocrine disruption is a serious concern. Therefore, there is a great need to pay attention to how food can be prevented from encountering artificial chemicals through new research. EDCs found in foods that cause undesirable health effects should be identified [120].

Since no clear and promising results have been obtained from the currently available scientific tests, further work in this area needs to be done to better understand the potential problems related to their use, and to reduce the risk of their exposure.

CONSENT FOR PUBLICATION

Not Applicable.

CONFLICT OF INTEREST

The author confirms that this chapter contents have no conflict of interest.

ACKNOWLEDGEMENT

Declared none.

REFERENCES

[1]   Bovee TFH, Schoonen WGEJ, Hamers ARM, Bento MJ, Peijnenburg AACM. Screening of synthetic and plant-derived compounds for (anti)estrogenic and (anti)androgenic activities. Anal Bioanal Chem 2008; 390(4): 1111-9.
[http://dx.doi.org/10.1007/s00216-007-1772-3] [PMID: 18188547]

[2]   Angeloni C, Teti G, Barbalace MC, Malaguti M, Falconi M, Hrelia S. 17β-Estradiol enhances sulforaphane cardioprotection against oxidative stress. J Nutr Biochem 2017; 42: 26-36.
[http://dx.doi.org/10.1016/j.jnutbio.2016.12.017] [PMID: 28110122]

[3]   Shang Y. Molecular mechanisms of oestrogen and SERMs in endometrial carcinogenesis. Nat Rev Cancer 2006; 6(5): 360-8.
[http://dx.doi.org/10.1038/nrc1879] [PMID: 16633364]

[4]   IUPAC-IUB Joint Commission on Biochemical Nomenclature (JCBN). The nomenclature of steroids. Recommendations 1989. Eur J Biochem 1989; 186(3): 429-58.
[http://dx.doi.org/10.1111/j.1432-1033.1989.tb15228.x] [PMID: 2606099]

[5]   London, New York: Chapman & Hall 1991.

[6]   Baker ME. What are the physiological estrogens? Steroids 2013; 78(3): 337-40.
        [http://dx.doi.org/10.1016/j.steroids.2012.12.011] [PMID: 23313336]

[7]   Hanukoglu I. Steroidogenic enzymes: structure, function, and role in regulation of steroid hormone
        biosynthesis. J Steroid Biochem Mol Biol 1992; 43(8): 779-804.
        [http://dx.doi.org/10.1016/0960-0760(92)90307-5] [PMID: 22217824]

[8]   Rossier MF. T channels and steroid biosynthesis: in search of a link with mitochondria. Cell Calcium
        2006; 40(2): 155-64.
        [http://dx.doi.org/10.1016/j.ceca.2006.04.020] [PMID: 16759697]

[9]   Lubik AA, Nouri M, Truong S, *et al.* Paracrine sonic hedgehog signaling contributes significantly to
        acquired steroidogenesis in the prostate tumor microenvironment. Int J Cancer 2017; 140(2): 358-69.
        [http://dx.doi.org/10.1002/ijc.30450] [PMID: 27672740]

[10]   Bhavnani BR, Nisker JA, Martin J, Aletebi F, Watson L, Milne JK. Comparison of pharmacokinetics
        of a conjugated equine estrogen preparation (premarin) and a synthetic mixture of estrogens (C.E.S.)
        in postmenopausal women. J Soc Gynecol Investig 2000; 7(3): 175-83.
        [http://dx.doi.org/10.1177/107155760000700307] [PMID: 10865186]

[11]   Liu M-J, Wang Z, Ju Y, Wong RN-S, Wu Q-Y. Diosgenin induces cell cycle arrest and apoptosis in
        human leukemia K562 cells with the disruption of Ca^{2+} homeostasis. Cancer Chemother Pharmacol
        2005; 55(1): 79-90.
        [http://dx.doi.org/10.1007/s00280-004-0849-3] [PMID: 15372201]

[12]   Nazari E, Suja F. Effects of 17β-estradiol (E2) on aqueous organisms and its treatment problem: a
        review. Rev Environ Health 2016; 31(4): 465-91.
        [http://dx.doi.org/10.1515/reveh-2016-0040] [PMID: 27883330]

[13]   Santen RJ, Kagan R, Altomare CJ, Komm B, Mirkin S, Taylor HS. Current and evolving approaches
        to individualizing estrogen receptor-based therapy for menopausal women. J Clin Endocrinol Metab
        2014; 99(3): 733-47.
        [http://dx.doi.org/10.1210/jc.2013-3680] [PMID: 24423357]

[14]   Gupta A, Ahmad I, Kureel J, *et al.* Differentiation of skeletal osteogenic progenitor cells to osteoblasts
        with 3,4-diarylbenzopyran based amide derivatives: Novel osteogenic agents. Eur J Med Chem 2016;
        121: 82-99.
        [http://dx.doi.org/10.1016/j.ejmech.2016.05.023] [PMID: 27236065]

[15]   Adsule S, Banerjee S, Ahmed F, Padhye S, Sarkar FH. Hybrid anticancer agents: isothiocyanate-
        progesterone conjugates as chemotherapeutic agents and insights into their cytotoxicities. Bioorg Med
        Chem Lett 2010; 20(3): 1247-51.
        [http://dx.doi.org/10.1016/j.bmcl.2009.11.128] [PMID: 20022750]

[16]   Scott SM, Brown M, Come SE. Emerging data on the efficacy and safety of fulvestrant, a unique
        antiestrogen therapy for advanced breast cancer. Expert Opin Drug Saf 2011; 10(5): 819-26.
        [http://dx.doi.org/10.1517/14740338.2011.595560] [PMID: 21699443]

[17]   Lee CI, Goodwin A, Wilcken N. Fulvestrant for hormone-sensitive metastatic breast cancer. Cochrane
        Database Syst Rev 2017; 1CD011093
        [PMID: 28043088]

[18]   Ashby J, Odum J, Paton D, Lefevre PA, Beresford N, Sumpter JP. Re-evaluation of the first synthetic
        estrogen, 1-keto-1,2,3, 4-tetrahydrophenanthrene, and bisphenol A, using both the ovariectomised rat
        model used in 1933 and additional assays. Toxicol Lett 2000; 115(3): 231-8.
        [http://dx.doi.org/10.1016/S0378-4274(00)00198-3] [PMID: 10814893]

[19]   Pavlik EJ. Estrogens, Progestins, and Their Antagonists: Health Issues. Boston, MA: Birkhäuser
        Boston 1996.

[20]   Chen X, Uzuner U, Li M, Shi W, Yuan JS, Dai SY. Phytoestrogens and Mycoestrogens Induce Signature Structure Dynamics Changes on Estrogen Receptor α. Int J Environ Res Public Health 2016; 13(9)E869
[http://dx.doi.org/10.3390/ijerph13090869] [PMID: 27589781]

[21]   Elks J, Ganellin CR. Dictionary of Drugs: Chemical Data. Structures, and Bibliographies 1990.
[http://dx.doi.org/10.1007/978-1-4757-2085-3]

[22]   Agarwal OP. Chemistry of Organic Natural Products V2. Meerut: Goel Publishing House 2006.

[23]   Medjakovic S, Jungbauer A. Red clover isoflavones biochanin A and formononetin are potent ligands of the human aryl hydrocarbon receptor. J Steroid Biochem Mol Biol 2008; 108(1-2): 171-7.
[http://dx.doi.org/10.1016/j.jsbmb.2007.10.001] [PMID: 18060767]

[24]   de Lemos ML. Effects of soy phytoestrogens genistein and daidzein on breast cancer growth. Ann Pharmacother 2001; 35(9): 1118-21.
[http://dx.doi.org/10.1345/aph.10257] [PMID: 11573864]

[25]   Vejdovszky K, Schmidt V, Warth B, Marko D. Combinatory Estrogenic Effects between the Isoflavone Genistein and the Mycotoxins Zearalenone and Alternariol in Vitro. Mol Nutr Food Res 2016.
[PMID: 27739238]

[26]   Rizzato G, Scalabrin E, Radaelli M, Capodaglio G, Piccolo O. A new exploration of licorice metabolome. Food Chem 2017; 221: 959-68.
[http://dx.doi.org/10.1016/j.foodchem.2016.11.068] [PMID: 27979300]

[27]   Athar M, Back JH, Tang X, *et al.* Resveratrol: a review of preclinical studies for human cancer prevention. Toxicol Appl Pharmacol 2007; 224(3): 274-83.
[http://dx.doi.org/10.1016/j.taap.2006.12.025] [PMID: 17306316]

[28]   Cos P, De Bruyne T, Apers S, Vanden Berghe D, Pieters L, Vlietinck AJ. Phytoestrogens: recent developments. Planta Med 2003; 69(7): 589-99.
[http://dx.doi.org/10.1055/s-2003-41122] [PMID: 12898412]

[29]   Dong Y-H, Wei J-H, Li Z. [The function of ERα in male reproductive system]. Sheng Li Ke Xue Jin Zhan 2014; 45(6): 410-5. [The function of ERα in male reproductive system].
[PMID: 25872345]

[30]   Valladares M, Plaza-Parrochia F, Lépez M, *et al.* Effect of estradiol on the expression of angiogenic factors in epithelial ovarian cancer. Histol Histopathol 2017; 32(11): 1187-96.
[PMID: 28116735]

[31]   Caballero I, Boyd J, Almiñana C, *et al.* Understanding the dynamics of Toll-like Receptor 5 response to flagellin and its regulation by estradiol. Sci Rep 2017; 7: 40981.
[http://dx.doi.org/10.1038/srep40981] [PMID: 28112187]

[32]   Wang Y-C, Xiao X-L, Li N, *et al.* Oestrogen inhibits BMP4-induced BMP4 expression in cardiomyocytes: a potential mechanism of oestrogen-mediated protection against cardiac hypertrophy. Br J Pharmacol 2015; 172(23): 5586-95.
[http://dx.doi.org/10.1111/bph.12983] [PMID: 25323043]

[33]   Speroni L, Voutilainen M, Mikkola ML, *et al.* New insights into fetal mammary gland morphogenesis: differential effects of natural and environmental estrogens. Sci Rep 2017; 7: 40806.
[http://dx.doi.org/10.1038/srep40806] [PMID: 28102330]

[34]   Asiedu B, Anang Y, Nyarko A, *et al.* The Role of Sex Steroid Hormones in Benign Prostatic Hyperplasia 2017.
[http://dx.doi.org/10.1080/13685538.2016.1272101]

[35]   Brisken C, O'Malley B. Hormone action in the mammary gland. Cold Spring Harb Perspect Biol 2010; 2(12)a003178

[http://dx.doi.org/10.1101/cshperspect.a003178] [PMID: 20739412]

[36]   Blackburn ST. Maternal, Fetal, & Neonatal Physiology: A Clinical Perspective. Amsterdam: Elsevier Saunders 2013.

[37]   Rosano GM, Panina G. Oestrogens and the heart. Therapie 1999; 54(3): 381-5.
        [PMID: 10500455]

[38]   Walter P, Green S, Greene G, *et al.* Cloning of the human estrogen receptor cDNA. Proc Natl Acad Sci USA 1985; 82(23): 7889-93.
        [http://dx.doi.org/10.1073/pnas.82.23.7889] [PMID: 3865204]

[39]   Weihua Z, Saji S, Mäkinen S, *et al.* Estrogen receptor (ER) beta, a modulator of ERalpha in the uterus. Proc Natl Acad Sci USA 2000; 97(11): 5936-41.
        [http://dx.doi.org/10.1073/pnas.97.11.5936] [PMID: 10823946]

[40]   Niu A-Q, Xie L-J, Wang H, Zhu B, Wang S-Q. Prediction of selective estrogen receptor beta agonist using open data and machine learning approach. Drug Des Devel Ther 2016; 10: 2323-31.
        [http://dx.doi.org/10.2147/DDDT.S110603] [PMID: 27486309]

[41]   Kiyama R, Wada-Kiyama Y. Estrogenic endocrine disruptors: Molecular mechanisms of action. Environ Int 2015; 83: 11-40.
        [http://dx.doi.org/10.1016/j.envint.2015.05.012] [PMID: 26073844]

[42]   Miller KKM, Al-Rayyan N, Ivanova MM, *et al.* DHEA metabolites activate estrogen receptors alpha and beta. Steroids 2013; 78(1): 15-25.
        [http://dx.doi.org/10.1016/j.steroids.2012.10.002] [PMID: 23123738]

[43]   Farzaneh S, Zarghi A. Estrogen Receptor Ligands: A Review (2013-2015). Sci Pharm 2016; 84(3): 409-27.
        [http://dx.doi.org/10.3390/scipharm84030409] [PMID: 28117309]

[44]   Barrett Mueller K, Lu Q, Mohammad NN, *et al.* Estrogen receptor inhibits mineralocorticoid receptor transcriptional regulatory function. Endocrinology 2014; 155(11): 4461-72.
        [http://dx.doi.org/10.1210/en.2014-1270] [PMID: 25051445]

[45]   Yi P, Wang Z, Feng Q, *et al.* Structure of a biologically active estrogen receptor-coactivator complex on DNA. Mol Cell 2015; 57(6): 1047-58.
        [http://dx.doi.org/10.1016/j.molcel.2015.01.025] [PMID: 25728767]

[46]   Martin TM. Prediction of *in vitro* and *in vivo* oestrogen receptor activity using hierarchical clustering. SAR QSAR Environ Res 2016; 27(1): 17-30.
        [http://dx.doi.org/10.1080/1062936X.2015.1125945] [PMID: 26784454]

[47]   Fratev F. Activation helix orientation of the estrogen receptor is mediated by receptor dimerization: evidence from molecular dynamics simulations. Phys Chem Chem Phys 2015; 17(20): 13403-20.
        [http://dx.doi.org/10.1039/C5CP00327J] [PMID: 25927714]

[48]   Hilder TA, Hodgkiss JM. Molecular Mechanism of Binding between 17β-Estradiol and DNA. Comput Struct Biotechnol J 2016; 15: 91-7.
        [http://dx.doi.org/10.1016/j.csbj.2016.12.001] [PMID: 28066533]

[49]   DeMayo F J, Zhao B, Takamoto N, Tsai S Y. Mechanisms of Action of Estrogen and Progesterone 2002.
        [http://dx.doi.org/10.1111/j.1749-6632.2002.tb02765.x]

[50]   Life: The Science of Biology . 9th ed., D. E. Sadava, Ed. Sinauer Associates ; W. H. Freeman : Sunderland, Mass. : Gordonsville, Va 2011.

[51]   Stormshak F, Bishop CV. Board-invited review: Estrogen and progesterone signaling: genomic and nongenomic actions in domestic ruminants. J Anim Sci 2008; 86(2): 299-315.
        [http://dx.doi.org/10.2527/jas.2007-0489] [PMID: 17965328]

[52]   Özen S, Darcan Ş. Effects of environmental endocrine disruptors on pubertal development. J Clin Res

Pediatr Endocrinol 2011; 3(1): 1-6.
[http://dx.doi.org/10.4274/jcrpe.v3i1.01] [PMID: 21448326]

[53] Roy JR, Chakraborty S, Chakraborty TR. Estrogen-like endocrine disrupting chemicals affecting puberty in humans--a review. Med Sci Monit 2009; 15(6): RA137-45.
[PMID: 19478717]

[54] Massart F, Parrino R, Seppia P, Federico G, Saggese G. How do environmental estrogen disruptors induce precocious puberty? Minerva Pediatr 2006; 58(3): 247-54.
[PMID: 16832329]

[55] Montes-Grajales D, Martínez-Romero E, Olivero-Verbel J. Phytoestrogens and mycoestrogens interacting with breast cancer proteins. Steroids 2018; 134: 9-15.
[http://dx.doi.org/10.1016/j.steroids.2018.03.010] [PMID: 29608946]

[56] Bingham SA, Atkinson C, Liggins J, Bluck L, Coward A. Phyto-oestrogens: where are we now? Br J Nutr 1998; 79(5): 393-406.
[http://dx.doi.org/10.1079/BJN19980068] [PMID: 9682657]

[57] Paterni I, Granchi C, Minutolo F. Risks and benefits related to alimentary exposure to xenoestrogens. Crit Rev Food Sci Nutr 2017; 57(16): 3384-404.
[http://dx.doi.org/10.1080/10408398.2015.1126547] [PMID: 26744831]

[58] Warth B, Preindl K, Manser P, Wick P, Marko D, Buerki-Thurnherr T. Transfer and Metabolism of the Xenoestrogen Zearalenone in Human Perfused Placenta. Environ Health Perspect 2019; 127(10)107004
[http://dx.doi.org/10.1289/EHP4860] [PMID: 31596610]

[59] Braun D, Ezekiel CN, Marko D, Warth B. Exposure to Mycotoxin-Mixtures via Breast Milk: An Ultra-Sensitive LC-MS/MS Biomonitoring Approach. Front Chem 2019; 8: 423.
[http://dx.doi.org/10.3389/fchem.2020.00423]

[60] Nebesio TD, Pescovitz OH. The Role of Endocrine Disruptors in Pubertal Development.When Puberty is Precocious. Totowa, NJ: Humana Press 2007; pp. 425-42.
[http://dx.doi.org/10.1007/978-1-59745-499-5_20]

[61] Jacobson-Dickman E, Lee MM. The influence of endocrine disruptors on pubertal timing. Curr Opin Endocrinol Diabetes Obes 2009; 16(1): 25-30.
[http://dx.doi.org/10.1097/MED.0b013e328320d560] [PMID: 19115521]

[62] Den Hond E, Schoeters G. Endocrine Disrupters and Human Puberty 2006.
[http://dx.doi.org/10.1111/j.1365-2605.2005.00561.x]

[63] Abaci A, Demir K, Bober E, Buyukgebiz A. Endocrine disrupters - with special emphasis on sexual development. Pediatr Endocrinol Rev 2009; 6(4): 464-75.
[PMID: 19550381]

[64] Diamanti-Kandarakis E, Bourguignon J-P, Giudice LC, et al. Endocrine-disrupting chemicals: an Endocrine Society scientific statement. Endocr Rev 2009; 30(4): 293-342.
[http://dx.doi.org/10.1210/er.2009-0002] [PMID: 19502515]

[65] Fénichel P, Chevalier N. Environmental endocrine disruptors: New diabetogens? C R Biol 2017; 340(9-10): 446-52.
[http://dx.doi.org/10.1016/j.crvi.2017.07.003] [PMID: 28826789]

[66] Vajda AM, Barber LB, Gray JL, Lopez EM, Woodling JD, Norris DO. Reproductive disruption in fish downstream from an estrogenic wastewater effluent. Environ Sci Technol 2008; 42(9): 3407-14.
[http://dx.doi.org/10.1021/es0720661] [PMID: 18522126]

[67] Skibińska I, Jendraszak M, Borysiak K, Jędrzejczak P, Kotwicka M. 17β-estradiol and xenoestrogens reveal synergistic effect on mitochondria of human sperm. Ginekol Pol 2016; 87(5): 360-6.
[http://dx.doi.org/10.5603/GP.2016.0005] [PMID: 27304652]

[68] Snoj T, Majdič G. MECHANISMS IN ENDOCRINOLOGY: Estrogens in consumer milk: is there a risk to human reproductive health? Eur J Endocrinol 2018; 179(6): R275-86.
[http://dx.doi.org/10.1530/EJE-18-0591] [PMID: 30400018]

[69] Singleton DW, Khan SA. Xenoestrogen exposure and mechanisms of endocrine disruption. Front Biosci 2003; 8: s110-8.
[http://dx.doi.org/10.2741/1010] [PMID: 12456297]

[70] Lloyd V, Morse M, Purakal B, *et al.* Hormone-Like Effects of Bisphenol A on p53 and Estrogen Receptor Alpha in Breast Cancer Cells. Biores Open Access 2019; 8(1): 169-84.
[http://dx.doi.org/10.1089/biores.2018.0048] [PMID: 31681507]

[71] Xu Z, Liu J, Wu X, Huang B, Pan X. Nonmonotonic responses to low doses of xenoestrogens: A review. Environ Res 2017; 155: 199-207.
[http://dx.doi.org/10.1016/j.envres.2017.02.018] [PMID: 28231547]

[72] Prossnitz ER, Barton M. The G-protein-coupled estrogen receptor GPER in health and disease. Nat Rev Endocrinol 2011; 7(12): 715-26.
[http://dx.doi.org/10.1038/nrendo.2011.122] [PMID: 21844907]

[73] Sidorkiewicz I, Zaręba K, Wołczyński S, Czerniecki J. Endocrine-disrupting chemicals-Mechanisms of action on male reproductive system. Toxicol Ind Health 2017; 33(7): 601-9.
[http://dx.doi.org/10.1177/0748233717695160] [PMID: 28464759]

[74] Brouwer A, Morse DC, Lans MC, *et al.* Interactions of persistent environmental organohalogens with the thyroid hormone system: mechanisms and possible consequences for animal and human health. Toxicol Ind Health 1998; 14(1-2): 59-84.
[http://dx.doi.org/10.1177/074823379801400107] [PMID: 9460170]

[75] Gray LE Jr, Ostby J, Cooper RL, Kelce WR. The estrogenic and antiandrogenic pesticide methoxychlor alters the reproductive tract and behavior without affecting pituitary size or LH and prolactin secretion in male rats. Toxicol Ind Health 1999; 15(1-2): 37-47.
[http://dx.doi.org/10.1191/074823399678846655] [PMID: 10188190]

[76] Kelce WR, Stone CR, Laws SC, Gray LE, Kemppainen JA, Wilson EM. Persistent DDT metabolite p,p′-DDE is a potent androgen receptor antagonist. Nature 1995; 375(6532): 581-5.
[http://dx.doi.org/10.1038/375581a0] [PMID: 7791873]

[77] Bigsby R, Chapin RE, Daston GP, *et al.* Evaluating the effects of endocrine disruptors on endocrine function during development. Environ Health Perspect 1999; 107 (Suppl. 4): 613-8.
[PMID: 10421771]

[78] Viñas R, Jeng Y-J, Watson CS. Non-genomic effects of xenoestrogen mixtures. Int J Environ Res Public Health 2012; 9(8): 2694-714.
[http://dx.doi.org/10.3390/ijerph9082694] [PMID: 23066391]

[79] Mantovani A, Maranghi F, La Rocca C, Tiboni GM, Clementi M. The role of toxicology to characterize biomarkers for agrochemicals with potential endocrine activities. Reprod Toxicol 2008; 26(1): 1-7.
[http://dx.doi.org/10.1016/j.reprotox.2008.05.063] [PMID: 18621504]

[80] Bretveld RW, Thomas CMG, Scheepers PTJ, Zielhuis GA, Roeleveld N. Pesticide exposure: the hormonal function of the female reproductive system disrupted? Reprod Biol Endocrinol 2006; 4: 30.
[http://dx.doi.org/10.1186/1477-7827-4-30] [PMID: 16737536]

[81] Wittliff JL, Andres SA. Estrogens V.Encyclopedia of Toxicology. Elsevier 2014; pp. 480-4.
[http://dx.doi.org/10.1016/B978-0-12-386454-3.01018-6]

[82] De Coster S, van Larebeke N. Endocrine-disrupting chemicals: associated disorders and mechanisms of action. J Environ Public Health 2012; 2012713696
[http://dx.doi.org/10.1155/2012/713696] [PMID: 22991565]

[83] Kucińska M, Murias M. [Cosmetics as source of xenoestrogens exposure]. Przegl Lek 2013; 70(8): 647-51. [Cosmetics as source of xenoestrogens exposure].
[PMID: 24466711]

[84] MacKenna C. Natural Breast Enlargement: The Ultimate Guide to Bigger, Firmer Breasts; Lulu.com: Place of publication not identified 2013.

[85] Wright-Walters M, Volz C. Municipal Wastewater Concentrations of Pharmaceutical and Xeno-Estrogens: Wildlife and Human Health Implications. In: Nzewi E, Reddy G, Luster-Teasley S, *et al.,* Eds., Proceedings of the 2007 National Conference on Environmental Science and Technology. 103-13.
[http://dx.doi.org/10.1007/978-0-387-88483-7_15]

[86] Acconcia F, Fiocchetti M, Marino M. Xenoestrogen regulation of ERα/ERβ balance in hormone-associated cancers. Mol Cell Endocrinol 2017; 457: 3-12.
[http://dx.doi.org/10.1016/j.mce.2016.10.033] [PMID: 27816767]

[87] Pastor-Barriuso R, Fernández MF, Castaño-Vinyals G, *et al.* Total Effective Xenoestrogen Burden in Serum Samples and Risk for Breast Cancer in a Population-Based Multicase-Control Study in Spain. Environ Health Perspect 2016; 124(10): 1575-82.
[http://dx.doi.org/10.1289/EHP157] [PMID: 27203080]

[88] Rosenfeld CS. Effects of Phytoestrogens on the Developing Brain, Gut Microbiota, and Risk for Neurobehavioral Disorders. Front Nutr 2019; 6: 142.
[http://dx.doi.org/10.3389/fnut.2019.00142] [PMID: 31555657]

[89] Poursafa P, Ataei E, Kelishadi R. A systematic review on the effects of environmental exposure to some organohalogens and phthalates on early puberty. J Res Med Sci 2015; 20(6): 613-8.
[http://dx.doi.org/10.4103/1735-1995.165971] [PMID: 26600838]

[90] de Cock M, van de Bor M. Obesogenic effects of endocrine disruptors, what do we know from animal and human studies? Environ Int 2014; 70: 15-24.
[http://dx.doi.org/10.1016/j.envint.2014.04.022] [PMID: 24879368]

[91] Mueller NT, Duncan BB, Barreto SM, *et al.* Earlier age at menarche is associated with higher diabetes risk and cardiometabolic disease risk factors in Brazilian adults: Brazilian Longitudinal Study of Adult Health (ELSA-Brasil). Cardiovasc Diabetol 2014; 13: 22.
[http://dx.doi.org/10.1186/1475-2840-13-22] [PMID: 24438044]

[92] Laessig SA, McCarthy MM, Silbergeld EK. Neurotoxic effects of endocrine disruptors. Curr Opin Neurol 1999; 12(6): 745-51.
[http://dx.doi.org/10.1097/00019052-199912000-00015] [PMID: 10676759]

[93] Gouva L, Tsatsoulis A. The role of estrogens in cardiovascular disease in the aftermath of clinical trials. Hormones (Athens) 2004; 3(3): 171-83.
[http://dx.doi.org/10.14310/horm.2002.11124] [PMID: 16982590]

[94] Urriola-Muñoz P, Li X, Maretzky T, *et al.* The xenoestrogens biphenol-A and nonylphenol differentially regulate metalloprotease-mediated shedding of EGFR ligands. J Cell Physiol 2018; 233(3): 2247-56.
[http://dx.doi.org/10.1002/jcp.26097] [PMID: 28703301]

[95] Shafei A, Ramzy MM, Hegazy AI, *et al.* The molecular mechanisms of action of the endocrine disrupting chemical bisphenol A in the development of cancer. Gene 2018; 647: 235-43.
[http://dx.doi.org/10.1016/j.gene.2018.01.016] [PMID: 29317319]

[96] Sheikh IA, Tayubi IA, Ahmad E, *et al.* Computational insights into the molecular interactions of environmental xenoestrogens 4-tert-octylphenol, 4-nonylphenol, bisphenol A (BPA), and BPA metabolite, 4-methyl-2, 4-bis (4-hydroxyphenyl) pent-1-ene (MBP) with human sex hormone-binding globulin. Ecotoxicol Environ Saf 2017; 135: 284-91.
[http://dx.doi.org/10.1016/j.ecoenv.2016.10.005] [PMID: 27750096]

[97]   Agas D, Lacava G, Sabbieti MG. Bone and bone marrow disruption by endocrine-active substances. J Cell Physiol 2018; 234(1): 192-213.
[http://dx.doi.org/10.1002/jcp.26837] [PMID: 29953590]

[98]   Ribeiro E, Ladeira C, Viegas S. Occupational Exposure to Bisphenol A (BPA): A Reality That Still Needs to Be Unveiled. Toxics 2017; 5(3)E22
[http://dx.doi.org/10.3390/toxics5030022] [PMID: 29051454]

[99]   Di Lorenzo M, Forte M, Valiante S, Laforgia V, De Falco M. Interference of dibutylphthalate on human prostate cell viability. Ecotoxicol Environ Saf 2018; 147: 565-73.
[http://dx.doi.org/10.1016/j.ecoenv.2017.09.030] [PMID: 28918339]

[100]   Bronowicka-Kłys DE, Lianeri M, Jagodziński PP. The role and impact of estrogens and xenoestrogen on the development of cervical cancer. Biomed Pharmacother 2016; 84: 1945-53.
[http://dx.doi.org/10.1016/j.biopha.2016.11.007] [PMID: 27863841]

[101]   Jurkowska K, Kratz EM, Sawicka E, Piwowar A. The impact of metalloestrogens on the physiology of male reproductive health as a current problem of the XXI century. J Physiol Pharmacol 2019; 70(3)
[PMID: 31539881]

[102]   Kresovich JK, Erdal S, Chen HY, Gann PH, Argos M, Rauscher GH. Metallic air pollutants and breast cancer heterogeneity. Environ Res 2019; 177108639
[http://dx.doi.org/10.1016/j.envres.2019.108639] [PMID: 31419716]

[103]   Forte M, Di Lorenzo M, Iachetta G, Mita DG, Laforgia V, De Falco M. Nonylphenol acts on prostate adenocarcinoma cells via estrogen molecular pathways. Ecotoxicol Environ Saf 2019; 180: 412-9.
[http://dx.doi.org/10.1016/j.ecoenv.2019.05.035] [PMID: 31108418]

[104]   Lillo MA, Nichols C, Perry C, *et al.* Methylparaben stimulates tumor initiating cells in ER+ breast cancer models. J Appl Toxicol 2017; 37(4): 417-25.
[http://dx.doi.org/10.1002/jat.3374] [PMID: 27581495]

[105]   Nowak K, Ratajczak-Wrona W, Górska M, Jabłońska E. Parabens and their effects on the endocrine system. Mol Cell Endocrinol 2018; 474: 238-51.
[http://dx.doi.org/10.1016/j.mce.2018.03.014] [PMID: 29596967]

[106]   Subramaniam K, Solomon J. Organochlorine pesticides BHC and DDE in human blood in and around Madurai, India. Indian J Clin Biochem 2006; 21(2): 169-72.
[http://dx.doi.org/10.1007/BF02912936] [PMID: 23105638]

[107]   Kim K-H, Kabir E, Jahan SA. Exposure to pesticides and the associated human health effects. Sci Total Environ 2017; 575: 525-35.
[http://dx.doi.org/10.1016/j.scitotenv.2016.09.009] [PMID: 27614863]

[108]   Konieczna A, Rutkowska A, Rachoń D. Health risk of exposure to Bisphenol A (BPA). Rocz Panstw Zakl Hig 2015; 66(1): 5-11.
[PMID: 25813067]

[109]   Almeida S, Raposo A, Almeida-González M, Carrascosa C, Bisphenol A. Food Exposure and Impact on Human Health: Bisphenol A and human health effect…. Compr Rev Food Sci Food Saf 2018; 17: 1503-17.
[http://dx.doi.org/10.1111/1541-4337.12388] [PMID: 33350146]

[110]   Rowdhwal SSS, Chen J. Toxic Effects of Di-2-ethylhexyl Phthalate: An Overview. BioMed Res Int 2018; 20181750368
[http://dx.doi.org/10.1155/2018/1750368] [PMID: 29682520]

[111]   Tchounwou PB, Yedjou CG, Patlolla AK, Sutton DJ. Heavy Metal Toxicity and the Environment.Molecular, Clinical and Environmental Toxicology. Basel: Springer Basel 2012; pp. 133-64. [Internet]
[http://dx.doi.org/10.1007/978-3-7643-8340-4_6]

[112] Jaishankar M, Tseten T, Anbalagan N, Mathew BB, Beeregowda KN. Toxicity, mechanism and health effects of some heavy metals. Interdiscip Toxicol 2014; 7(2): 60-72.
[http://dx.doi.org/10.2478/intox-2014-0009] [PMID: 26109881]

[113] Ghorani-Azam A, Riahi-Zanjani B, Balali-Mood M. Effects of air pollution on human health and practical measures for prevention in Iran. J Res Med Sci 2016; 21: 65.
[http://dx.doi.org/10.4103/1735-1995.189646] [PMID: 27904610]

[114] Endocrine Disruption and Human Health [Internet]. Elsevier; 2015 [cited 2020 Oct 23].

[115] Lauretta R, Sansone A, Sansone M, Romanelli F, Appetecchia M. Endocrine Disrupting Chemicals: Effects on Endocrine Glands. Front Endocrinol (Lausanne) 2019; 10: 178. [Internet].
[http://dx.doi.org/10.3389/fendo.2019.00178] [PMID: 30984107]

[116] Kahn LG, Philippat C, Nakayama SF, Slama R, Trasande L. Endocrine-disrupting chemicals: implications for human health. Lancet Diabetes Endocrinol 2020; 8(8): 703-18.
[http://dx.doi.org/10.1016/S2213-8587(20)30129-7] [PMID: 32707118]

[117] Kabir ER, Rahman MS, Rahman I. A review on endocrine disruptors and their possible impacts on human health. Environ Toxicol Pharmacol 2015; 40(1): 241-58.
[http://dx.doi.org/10.1016/j.etap.2015.06.009] [PMID: 26164742]

[118] Varjani S, Sudha MC. Occurrence and human health risk of micro-pollutants—A special focus on endocrine disruptor chemicals.Current Developments in Biotechnology and Bioengineering. Elsevier 2020; pp. 23-39. Internet
[http://dx.doi.org/10.1016/B978-0-12-819594-9.00002-4]

[119] FitzGerald RE. Perspective on Health Effects of Endocrine Disruptors with a Focus on Data Gaps. Chem Res Toxicol 2020; 33(6): 1284-91.
[http://dx.doi.org/10.1021/acs.chemrestox.9b00529] [PMID: 32250608]

[120] Rashid H, Alqahtani SS, Alshahrani S. Diet: A Source of Endocrine Disruptors. Endocr Metab Immune Disord Drug Targets 2020; 20(5): 633-45.
[http://dx.doi.org/10.2174/1871530319666191022100141] [PMID: 31642798]

CHAPTER 4

Persistent Organic Pollutants: The Ancient Intruders of Our Environment

Devendra Kumar Patel[1,*], **Neha Gupta**[1], **Sandeep Kumar**[1] and **Juhi Verma**[1]

[1] *CSIR-Indian Institute of Toxicology Research, Lucknow, India*

Abstract: Persistent organic pollutants (POPs) are chemicals compounds that directly affect human and animal health and accumulate in the environment leading to continuous exposure. They have the properties like bioaccumulation, persistence, and biomagnification, which give them the advantage of getting transported by wind and water. POPs were introduced with an intent to benefit the human population, but their excessive usage had made their presence everywhere that turns them to be toxic compounds. Some of the POPs are even generated as a byproduct of chemical and thermal processes. The toxicity evaluation of the POPs moved it towards the class of toxicants that are carcinogenic by nature and imposes a threat to animal and human both. The various analytical approaches had been made to quantify the POPs in various matrices using different sophisticated analytical tools like high-resolution GC-MS and LC-MS/MS. The limit of detection (LOD) and limit of quantification (LOQ) are proposed to be lower as they need to be quantified and detected in biological samples as well.

Keyword: Bioaccumulation, Biomagnification, Carcinogenic, High-resolution GC-MS and LC-MS/MS, Persistent.

INTRODUCTION

In the worldwide scenario, persistent organic pollutants (POPs) are responsible to cause adverse effects on the environment and human health due to their toxicity. The major issues associated with POPs are their easier transportation by both wind and water affecting people and wildlife in a wide range of areas. Therefore, it is difficult to locate the actual source of POPs contamination as it can be generated in one country and shows its effects even in an area far from where they are released. POPs are volatile compounds under environmental temperatures,

---

* **Corresponding author Devendra Kumar Patel:** Analytical Chemistry Laboratory & Regulatory Toxicology Group, CSIR-Indian Institute of Toxicology Research, Lucknow, India; E-mails: dkpatel@iitr.res.in and dkp27701@gmail.com

Tahmeena Khan, Abdul Rahman Khan, Saman Raza, Iqbal Azad and Alfred J. Lawrence (Eds.)

therefore gets volatilize from various sources they are being used in like soils, vegetation, and aquatic system. POPs resist the breakdown reactions in atmospheric air consequently and they travel to long distances before getting re-deposited and show their presence in regions far from where they were used or discharged. POPs can persist in the environment for a longer time and their bioaccumulation can be observed which leads to the transportation of POPs from one organism to another *via* the food chain. POPs can move through the food chain by getting accumulated in the body fat of living organisms and their concentration level increases at each tropic level that leads to "biomagnifications". Contaminants found in small concentrations at the lower level of the food chain can lead to biomagnifications thereby resulting in the catastrophe of the entire food web and therefore small release of POPs is responsible for remarkable results. This was confirmed by the analysis of the sample collected from the Antarctic region. The sample of melted glaciers shows a high concentration of POPs and proves the theory of distant migration of POPs through the air and concluded that melting glaciers are a secondary source of POPs [4].

Many sources like effluent waste releases, runoff from agricultural lands, and atmospheric deposition are responsible for the settlement of POPs in marine and freshwater ecosystems because of which these sediments act as reservoirs or "sinks" for POPs. Lower solubility in water made it possible for POPs to bond strongly to particulate matters in aquatic sediments. POPs have both hydrophobic (water repellant) and lipophilic (fat attracting) properties due to which they show strong binding capacity with solids, mainly organic matters, and can easily enter into the lipids of organisms and get stored in fatty tissues. This stockpiling of POPs in various types of fatty tissues makes them rich in these compounds and gets them preserved in the biota due to slow metabolism because of which, POPs move upwards in the food chain pyramid [2].

After World War II when the whole world was moving towards fast-pacing industrialization, the use of synthetic chemicals increased as they proved to be beneficial in pest and disease control to increase crop production and flourish the new industry. These chemicals besides providing a strategy for increasing crop production also put forth unforeseen effects on human health and the environment. With the increased risk of POPs exposure and harmful effect, an effort was made on a global level through the Stockholm Convection. The Convention has played an important role in combating harmful chemicals worldwide.

The major classification of POPs is:

1. Intentionally produced POPs: These chemicals were produced with a positive intention of using them in various fields according to their effectiveness. These chemicals include agricultural pesticides, disease control sprayer's reagents, and industrial important chemicals. Foe example: DDT is used as a mosquito repellent and PCBs are used in a variety of industrial applications. They are also used in the heat exchange and paint industry.
2. Unintentionally produced POPs: They include chemicals generated as byproducts of some other processes. These are not produced with any motive but are generated along with main compounds. For example: Dioxins are produced as a result of industrial processes and waste combustion.

Environmental Protection Agency (EPA) and the USA had notably lowered the release of POPs generated through combustion processes like dioxins and furans into the environment. Along with the assessment of dioxins like POPs, EPA had worked attentively on the reduction of DDT globally from primary sources. To reduce the emissions of POPs, different countries have signed an agreement to implement the elimination process of toxicants. The convention aims at reducing the generation of POPs, therefore in this context, the trade of POPs is prohibited to combat their production and use. The export of POPs is allowed under the convention only when the exporting countries certify that they would minimize the harmful release of POPs to the environment and attention would be paid to their destruction or disposition in an environment-friendly manner.

Through several studies conducted to assess the toxicity level of POPs, the findings showed abnormalities and decrease in the number of wildlife organisms such as certain kinds of fish, birds, and mammals, reproductive and developmental defects, behavioural changes, neurological, endocrine, and immunological problems in human beings. The sensitive groups comprising children and old aged people have a lower immune response that makes them more vulnerable to different POPs. They also cause reproductive toxicity in both men and women [1]. The possible route of exposure to POPs includes primarily contaminated food, whereas drinking water and direct contamination are lesser-known for POPs toxicity. The low levels of POPs exposure to humans are even dangerous because they can lead to increased chances of cancer, reproductive disorders, alteration in the immune response, degeneracy in neurobehavioral, disruption of endocrine functioning, genotoxicity and increased birth defects which are discussed in detail in the later part of the chapter.

To trace the body burdens of POPs, the matrices chosen are human milk, maternal blood, and adipose tissue. Biomonitoring of human milk gives detailed information about the exposure of the mother and the infants. Food is the major source of POPs exposure, WHO started the "GEMS/Food Programme" in 1976 in which detailed information about the level of POPs in food has been provided, including human milk. Further to clear the scenario regarding the level of POPs in human milk from 1987-2003, three international studies were commenced to assess the levels of dioxins, furans, and biphenyls (PCBs).

Dioxins being unintentionally generated toxins are a potential source for human health risk and are present as a complex mixture in the ecosystem. The most toxic compound of the dioxin family is TCDD which is used as a reference compound for assessing toxicity potential for all other dioxins. These values are based on experimental studies and 90% of human exposure is through food, mainly meat and dairy products, fish and shellfish worldwide [3].

Bioaccumulation and biomagnification of POPs in the food web result in deleterious effects on human health. As per the report established by analyzing the concentration of organic pollutants in the breast milk of French, Danish, and Finnish women, the result concluded that Danish women had a higher concentration of POPs in breast milk (1.5-2 times) than the French and finish women [5]. A similar study conducted by Long *et al.* with the Greenlandic women in Greenland also found that the concentration of POPs was higher in the blood of pregnant women due to the high intake of marine mammals. This study also confirmed the presence of POPs in many Arctic communities. At low temperature, the breakdown of POPs became very slow and they remained as such in the environment for a longer time [6]. Study on type 2 diabetes patients showed high levels of POPs (PCBs and OCPs) in the blood of women sufferers in Norway rather than the rest of the population, as reported by Rylander *et al.* Although the study did not prove that organic pollutants cause type II diabetes but gave a clue that organic pollutants may be responsible for the onset of type II diabetes [7].

ENVIRONMENTAL EFFECT OF POPS

Persistent organic pollutants (POPs) are usually referred to as "Silent killers" or "Forever chemicals" because they remain considerably unaffected by any type of degradation through any processes (biological, chemical, and photolytic). POPs are organic compounds that persist in nature for a longer time and have the potential for long-range transport, causing bioaccumulation and biomagnifications that lead to various adverse effects on the environment (Fig. **1**). As published under Stockholm convection, 2001 the mentioned POPs were regarded as a serious threat and needed to be combatted as soon as possible or the usage must have been reduced to the minimum. In 1960 a book titled 'Silent Spring' emphasised the use and effects of DDT. This book briefly described the declining cause of bird populations and ecosystem disruptions. There exists an intense relationship between air, water, living beings, and the environment. Any disturbance to any of the components can perturb the entire balance of the environment. Temperature is an essential component for the determination of the global distribution of POPs, as it directly affects the half-life, volatilisation, partitioning, and reemissions of POPs and indirectly by the formation of the hydroxyl radical formation process. Temperature also regulates the secondary emission of POPs from already contaminated environments by increasing its volatility, fast and rapid degradation, and make changes in partitioning between different phases. Several studies have been reported regarding the influence of climate change on the fate and transport of POPs to elucidate the effects on their long-range transport.

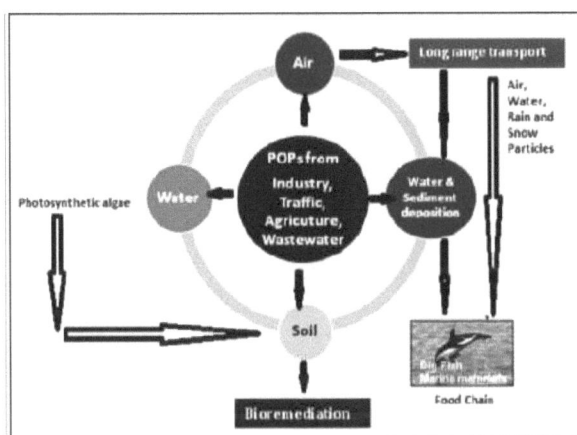

Fig. (1). Schematic depiction of the fate of POPs in the environment.

A brief description of the effect of environmental factors on the fate and transport of some of the POPs is as follows:

1. Pesticides

In Stockholm convection, pesticides form a large group of POPs. Several researchers studied the effect of climate change on pesticides. In 2010, Ma and Cao *et al*. [8] studied the quantification and perturbations of POPs through climate change. The selected pesticides for this study include α- and γ- HCHs, as well as HCB and PCB-153 congener showing high response to the specified climate change for perturbation in the air by 4-50% increase. An increase in the HCHs was observed in coastal sediments due to an increase in atmospheric temperature in the late 1970s. According to the analysis of water, performed from the coastal and non-coastal regions clearly depicted the difference that was caused due to the climate warming leading to the melting of glaciers and thus release of stored DDT into lakes [9]. Several other studies concluded that the phenomenon of re-volatilization of α-HCH, DTT and cis-chlordane that were deposited in water and ice sinks due to warming can be reduced by global efforts may minimise exposure [10]. In 2011, Hallanger *et al.* [11] assessed the differences in the bioaccumulation of POPs in zooplanktons. It was earlier reported that bioaccumulation leads to the occurrence of OCPs in the environment which can be biomagnified in the food chain. According to toxicology reports, these pesticides were carcinogenic and may create problems on their exposure to the human population. Dichloro diphenyl trichloroethane (DDT) and its derivatives and Hexachlorobenzene (HCB) can bioaccumulate in human adipose tissues.

2. Dioxins, Furans, and Polychlorinated Biphenyls

Polychlorinated dibenzo-p-dioxins (PCDDs) are also called Dioxins, while furans are polychlorinated dibenzo- furans (PCDFs). These are found in almost every part of the environment (like air, water, and soil) causing grave concerns because of their high toxicity towards humans, animals, and birds, *etc.* The increase in the use of chlorinated compounds and man-made activities (Chemical and combustion process, secondary sources), causes a steady increase in the concentration of PCDDs and PCDFs in the environment. In the 1970s, Polychlorinated biphenyls (PCBs) were banned by the developed nations, before that they were used widely as coolants, lubricants, and a paint additive. The main source of PCB pollution is the release into the air (industrial emission) during the manufacturing process and other pollution sources like weathering and incineration of PCB-containing materials, from defunct equipment and leaching from landfills. It was quite clear from the study conducted in the 1980s in China that the people residing around the industrial areas have higher dioxin levels in their adipose tissue which was 10-folds higher than those inhabiting rural China (0.142 ng/g). The effect of climate change was properly established by Chi *et al.*, 2013 [12] for their fate and transport that concludes that primary and secondary release of PCDD/Fs is directly affected by the climate change like higher wind speed which leads to the potential remobilization of previously deposited pollutants.

3. Polybrominated Diphenyl Ethers (PBDEs)

PBDEs are widely used as flame retardant for consumer plastics items, textiles, and electronic circuitry for consumer and industrial products and other materials. In 2001, the total worldwide demand for Deca-BDE, Octa-BDE, and Penta-BDE was 67 metric tons per year. In the last two decades, the production of PBDEs has significantly increased. It has been established that PBDEs are highly persistent, bio-accumulative, but their toxicological effect is under study to date.

4. Polycyclic Aromatic Hydrocarbons (PAHs)

Polycyclic aromatic hydrocarbons (PAHs) are a group of chemically related compounds that persist in the environment with variant structures and toxicity generated through many natural and anthropogenic activities in the environment. PAHs can easily be bioaccumulated in the environment and cause serious threats to the environment and various organisms. Therefore, the degradation rate of PAHs plays a vital role in controlling the PAHs concentration in soil, and water, *etc.*

HEALTH EFFECT OF POPS

POPs are a serious threat to human health because of their bioaccumulation in various tissues and non-degrading nature. POPs are exposed to humans primarily by the dietary intake of dairy products, fish, and meat, therefore diet plays an important role in the exposure of POPs. POPs have the unique property of being semi-volatile due to which they can travel miles before settling down. They are not readily metabolized or excreted due to their lipophilic nature and get easily accumulated in adipose tissues. In today's scenario, everyone is exposed to POPs in one or another way including the foetus and embryo. Several studies have been reported related to the working mechanism of developing problems and health effects due to POPs. In this context, Zhou *et al.* [13] and Mrema *et al.* [14] described the possible mechanism, which is complex in a model organism, zebra fish by aryl hydrocarbon receptor pathways and genomic and non-genomic pathways for the major OCPs that leads to diseases and health problems, respectively. Given below is a list of health effects caused by exposure to POPs:

Endocrine Disruptions

Endocrine disruptors are those chemicals that put forth a negative impact on the overall body system including reproduction, neurological and immune responses. The endocrine system is an essential component of the human body which includes various glands such as the pancreas, pituitary gland, adrenals, and testes that are directly linked to the human body growth factor. Many research works have been published regarding the endocrine disruption of POPs including exposure to the foetus [15 - 21]. In this context, organochlorine pesticides have been identified as a major class of endocrine disruptors [22]. They lead to various toxic neurological problems in children, impairing their overall development [22 -

25]. Even dioxins [25, 26] and PCBs [26 - 28] also lead to the neurological impairment observed in monkeys and rats' offspring.

Reproductive Defects

In today's scenario, reproductive disorders can be observed in every second female or male and it has become quite a prominent problem. Increasing the concentration of the POPs in the environment directly affects the reproductive health of humans by developing various types of reproductive diseases. The major ones are birth defects with developmental disorders, premature birth causing low body weight in neonatal, sterility leading to impotence, menstrual disorders which reduce the fertility rate in females [29 - 31]. Various researchers have reported the impact of various POPs on reproductive organs. Gao and Wu [29] and Damstra [15] reported the effect of POPs on male reproductive organs and according to their findings PCBs cause differential response in testes and decrease sperm quality and quantity, respectively. In females, exposure to POPs causes early puberty, affects pregnancy and endometriosis. Several studies showed the transfer of POPs from mother to offspring and related defects in newborns [32 - 37].

Cardiovascular Problems

The major cause of mortality in the world is cardiovascular problems which contribute to approximately 17 million deaths per year. Out of the total deaths, 9.4 million deaths are linked to hypertension [38 - 40]. POPs are one of the reasons for cardiovascular diseases like heart failure, cardiomyopathy hypertension, *etc.* Due to their lipophilic nature, they are bioaccumulated in the blood stream and cause various cardiac problems which are well supported by various authors in their reported researches [41 - 44].

Cancer

According to GLOBOCAN 2012, cancer was the cause of most of the deaths that occurred worldwide in 2012. It is also the second most lethal disease responsible for the deaths that occur annually. Cancer is a lethal disease that is caused when normal body cell growth is uncontrolled. Despite the advancement of medical sciences, yet the disease remains non-curable in the late and advanced stages. POPs are one of the reasons behind the cause of cancer as biomagnifications and bioaccumulation of high level of POPs in LDL (low-density lipoproteins) is responsible for various cancers as reported [45]. Yu *et al.* [46] in their study reported the risk of cancer by marine food consumption that act as sinks for POPs and people consuming the marine foods are much prone to cancer. Even breast cancer in females is excessively being reported these days and various studies have established the direct relationship between the concentration of POPs and breast cancer [47 - 49]. A study conducted in Jaipur, India by Mathur *et al.* [50] showed that pesticide exposure (DDT and its metabolites) and concentration in women suffering from breast cancer. Prostate cancer in males is also caused by POPs [51].

Diabetes

Diabetes is a metabolic syndrome caused due to insulin resistance that causes a spike in blood sugar levels. The acute symptoms of diabetes are diabetic acidosis and hyperosmolar hyper-glycemic nonketotic coma. Long-term co-morbidities associated with diabetes are heart attack, renal failure, frequent urination, ulcers in the foot, and blurred vision which is also caused due to bioaccumulation of POPs. Professor D.H. Lee did remarkable work in establishing a relationship between diabetes and POPs contamination, by showing the presence of POPs in serum and their effect on type II diabetes in different groups. Different researchers have studied the effect of organochlorine pesticides, PCB, and other POPs on diabetes [52 - 62].

Table 1. The main POPs involved in particular health problems.

Health Problems	Main POPs Responsible for Particular Health Problems
Breast, prostate, testicular, ovarian, uterine and kidney Cancer	PCBs, DDE, HCB, chlordane, pesticides, phthalates, PAHs, TeBDE and phytoestrogens
Type II diabetes	PCBs, DDT, DDE, HCB, PCDDs, PCDFs, 2,2',4,4',5,5'-Hexachlorobiphenyl (CB-153) and 1,1-dichloro-2,2-bis (p chlorophenyl)-ethylene
Cardiovascular problems	PCBs, OCDD, OCPs and one flame retardant brominated compound (BDE47)
Blood pressure	PCB congeners and Organochlorine pesticides
Endocrine disruptors	Oxychlordane, PCB, DDE, DDT, APEOs and PAHs
Reproductive defects	PCBs and DDTs

The Stockholm Convention, 2011 covered 12 POPs called the "dirty dozen" namely, aldrin, chlordane, DDT, dieldrin, dioxins, endrin, furans, heptachlor, and hexachlorobenzene, mirex, PCBs and toxaphene (Table **2 - 3**).

These toxic chemicals are still in use worldwide, although some of them are permanently banned due to their high rates of contamination in the environment such as Chlordecone and Hexabromobiphenyl. OctaBDE is highly persistent (makes it capable of long-range transport), bio accumulative, and biomagnified along the food chain. By-products of Lindane production are around 6-10 tons, such large stockpiles can lead to site contamination. DecaBD an additive flame retardant contains plastics. HCBD is bio accumulative and highly toxic for aquatic organisms and birds. PCBs are unintentionally produced chemical toxicants due to combustion and used in PCB products [64].

Table 2. The twelve persistent organic pollutants under the Stockholm Convention [63].

S. No.	Name of POPs	Use and Adverse Effects on Human & Ecosystem
1	**Aldrin and Dieldrin**	Uses: Aldrin and dieldrin are used to kill insecticides like termites, grasshoppers, corn rootworm, and other insect pests found in agricultural farmland soils. Aldrin is easily converted to dieldrin and leads to a high concentration of dieldrin in farmlands. Effects: Pesticides used in excess are harmful to birds, fish, aquatic organisms, and humans. Dieldrin exposure to aquatic organisms leads to embryos spinal deformities even at a low level of exposure.
2	**Chlordane**	Uses: Chlordane is mainly used for controlling broad-spectrum insecticides and termites that can cause harmful effects to agricultural croplands. Chlordane is responsible for killing mallard ducks (Anas platyrhynchos), bobwhite quail (Colinus virginianus), and pink shrimp (Pandalus Borealis). Effects: Chlordane is found to suppress the human immune system and causes cancer. Chlordane exposure to the human population is mainly through the air which is reported in a study conducted in the US and Japan wherein indoor air in households showed chlordane contamination.
3	**DDT**	Uses: DDT was used for various diseases like malaria and typhus during World War II which spread through insects. But its use was continued even after the war due to its lethal effect on pests in a variety of agricultural crops, especially cotton. Effects: Its stability and persistence made it possible to be found everywhere even in the region of the arctic. A highly used category of pesticide and its presence can be detected in breast milk, raising serious concerns about infant health.
4	**Endrin**	Uses: Endrin is used to kill rodenticides like mice and voles by spraying in cotton and grains crops. Endrin is easily metabolized by animals and therefore it does not bioaccumulate in adipose tissues. Effects: Endrin concentration in water shows early hatching of sheep head minnows which leads to early deaths and high toxicity for fish.
5	**Mirex**	Uses: Mirex is a pesticide that is used to protect the crops from destruction and to maintain high yield. Effects: As such, no reports are available for the direct exposure to mirex and its adverse effects on humans. On laboratory scale study it has been proven to be toxic to animals due to its carcinogenic activity.

(Table 2) cont.....

S. No.	Name of POPs	Use and Adverse Effects on Human & Ecosystem
6	**Heptachlor**	Uses: Heptachlor is responsible for killing soil insects and termites. It is widely used to kill crop pests, cotton insects, grasshoppers, and mosquitoes causing malaria. Effects: A study reported in the US about the Columbia River basin shows the cause of the decline in Canadian Geese and American Kestrels bird populations. The death was due to after eating seeds treated with heptachlor. Laboratory studies have shown that high doses of heptachlor are fatal for mink, rats, and rabbits. The lower dose causes adverse behavioral changes and reduced reproductive capacity.
7	**Hexachlorobenzene**	Uses: Use of Hexachlorobenzene leads to crop loss. It was introduced in 1945 for the treatment of seeds and control of wheat bunt. Effects: Traces of HCB were also found in mother milk showing its transportation through the placenta and breast milk, exposing the newborn baby directly to the HCB. HCB is harmful in both low and high doses as it leads to reproductive difficulties and animal lethality.
8	**PCBs**	Uses: PCBs are used basically as heat exchange fluids in industry, electric transformers, and capacitors. They are even added as additives in paint, carbonless copy paper, and plastics. Effects: PCBs are responsible for pigmentation of nails and mucous membranes, they even cause nausea leads to vomiting. Acute exposure leads to fatigue, and protuberance of the eyelids.
9	**Toxaphene**	Uses: Toxaphene is used as an insecticide to prevent cotton, cereal grains, fruits, nuts, and vegetables. The preservation of livestock from ticks and mites is also carried out with the help of toxaphene. Effects: Toxaphene has been listed as a possible source of human carcinogen.
10	**PCDFs**	Uses: As such the uses of Dioxins and Furans are not reported. These are mainly the byproducts of incomplete combustion of wastes that include hospital waste, municipal waste, and hazardous waste. They are even generated from automobile emissions, peat, coal, and wood. Effects: Dioxins cause adverse effects on humans, by impairing immune and enzyme responses. Both are classified as carcinogens and can cause cancer to human beings.

REMEDIATION

The existence of persistent organic pollutants in our environment is a global issue and a big challenge. They impose many mutagenic and carcinogenic effects on the food hierarchy and eventually on human health. Studies are required to understand the direct and indirect effects of POPs on animals, plants, human health, and their environmental impact. To eliminate the presence of POPs many techniques are introduced like thermal, chemical, and biological and they are also used in combinations. There are some environmental techniques also available for remediation of POPs like green nanotechnology and magnetic biochar. The application of remediation methods directly depends upon the cost-effectiveness, process time, and eco-friendliness [86] but further advancements are needed to diminish the use of these contaminated and hazardous substances from the environment [87]. POPs are harmful to both environment and human health and there is an urgent need to develop remedial methods for POPs. Usually, there are three ways by which these can be reduced:

• Reduced production and use of the POPs.
• Development of bioremediation methods for POPs.
• Removal of POPs through various chemical processes.

Table 3. The new POPs chemicals under the Stockholm Convention (2017) [64].

S.No.	Persistent Organic Pollutants
1.	Alpha hexachloro-cyclohexane (HCH)
2.	Beta hexachlorocyclo hexane
3.	Chlordecone

S.No.	Persistent Organic Pollutants
4.	 **Decabromodiphenyl ether**
5.	 **Hexabromobiphenyl**
6.	 **Hexabromocyclododecane**
7.	 **Hexabromodiphenyl ether and heptabromodiphenyl ether**
8.	 **Hexachlorobutadiene**

S.No.	Persistent Organic Pollutants
9.	Lindane
10.	Pentachlorobenzene
11.	Pentachlorophenol and its salt and esters (PCP)
12.	$CF_3(CF_2)_6CF_2SO_3^-$
13.	Polychlorinated napthalenes

S.No.	Persistent Organic Pollutants
14.	**Short chain chlorinated paraffin**
15.	**Technical endosulfan and its related Isomers**
16.	Tetrabromodiphenyl ether and pentabromodiphenyl ether (Commercial Penta bromodiphenyl ether)
17	**Perfluorooctanoic acid (PFOA), its salts and PFOA-related compounds**

POPs can enter the body through these routes: Skin Contact: Touching the products containing or made with POPs, Inhalation: Breathing contaminated air with these chemicals and Ingestion: Eating and drinking of food and water contaminated with these chemicals. Mostly these persistent chemicals can damage the immune system and ruins the natural mechanism of the neurological and reproductive system, endocrine disruption, decreased comprehension, greater susceptibility to disease, birth defects, cancer, learning disabilities, tumour generation, neurobehavioral deficiency, and growth deformities, *etc* [65]. Analytical researchers have analysed these chemicals in different matrices using sensitive extraction technique and state of art chromatographic techniques to achieve the lowest detection limit and lowest quantification limit. Tables **4** and **5** summarize the extraction techniques as well as analytical methods used by different researchers worldwide.

Table 4. Analytical instrument used in the analysis of pesticides with LODs.

Serial No.	Analyte	Instrument	Matrix	LODs	Reference
1.	Organ chlorines and Triazines	GC-ECD	Sugar cane juice	0.003-0.04 mg/L	[66]
2.	Multiclass pesticides	GC-ECD	Honey	0.03-10.6 ng/g	[67]
3.	Organophosphates	GC-FID	Water	0.82-2.72 ng/mL	[68]
4.	Organophosphates	GC-FPD	Water	0.043-0.085 μg/L	[69]
5.	Multiclass pesticides	GC-NPD	Fruit juice	0.05-0.43 ng/mL	[70]
6.	Organophosphates	GC-NPD	Fruit, vegetable and water	0.02-0.2 μg/L	[71]
7.	Organophosphates	GC-ICD/MS	Water	0.005-0.020 μg/L	[72]
8.	Organ chlorines	GC-MS/MS	Drinking and environmental water	0.39-0.40 ng/L	[73]
9.	Multiclass pesticides	GC-MS/MS	Artichoke leaves and fruits	0.005-0.025mg/Kg	[74]
10.	Carbamates	UPLC-MS/MS	Environmental Water	0.5-6.9 ng/L	[75]
11.	Multiclass pesticides	LC-ESI-MS/MS	Honey	0.35-7.09 ng/g	[76]
12.	Acaricides	HPLC-DAD	Fruit juice	0.16-0.57 μg/L	[77]
13.	Organophosphates	ELISA	Water	0.32 ng/mL	[78]

Table 5. Major analytical methods for the analysis of different POPs.

Serial No.	Analyte	Extraction Technique	Instrument	Reference
1.	PCDDs, PCDFs, PBDDs, PBDFs, DLPCBs, and PCNs	SPE	GC-HRMS	[79]
2.	PCBs and PBDEs	SPE	GC-MS	[80]
3.	PCDDs, PCDFs, and DLPCBs	SPE	GC-HRMS	[81]
4.	PCDDs, PCDFs, PBDEs, and OC pesticides	SPE	GC-HRMS and CZC-GC - HRMS	[82]
5.	PCBs, BFRs, and OC pesticides	SPE	GC-MS	[83]
6.	PCBs, BFRs, and OC pesticides	SPE	GCHRMS and GCxGC –TOF/MS	[84]
7.	PCBs, PBDEs, and OC pesticides	LLE	GC-MS and GC-ECD	[85]

(Table 5) cont.....

Serial No.	Analyte	Extraction Technique	Instrument	Reference
8.	PFAAs	SPE	LC –ESI MS/ MS	[86]
9.	PFAAs	LLE	HPLC –ESI MS/ MS	[87]
10.	OC pesticides	SPE	GC-HRMS	[88]

Microbial Degradation

Microbial degradation involves bioremediation and biotransformation, in which organic pollutants are removed from the environment (*i.e.*, Soil and Water) using micro-organisms. It is an economical and eco-friendly method that can be aerobic or anaerobic. These micro-organisms attack the basic structure of the compound or functional group which degrades by producing water, carbon dioxide, and salts. For highly chlorinated compounds, anaerobic degradation is preferred. The microbial degradation can be of the following types:

1. Bacterial Degradation

Bacteria are a class of prokaryotic single-celled organisms that are involved in the biodegradation/bioremediation of persistent pollutants. Micro-organism helps in maintaining the geochemical cycle, which leads to the sustainable development of the environment. Some examples of bacterial degradation are as follows:

a. *Novosphingobium*sp. PCY, *Microbacterium*sp. BPW, *Ralstonia sp.* BPH, *Alcaligenes sp.* SSK1B, and *Achromobacter sp.* SSK4 and PCY have degraded a large class of PAHs.
b. *Acinetobacter* sp, *Pseudomonas putida*, *Bacillus* sp, *Pseudomonas aeruginosa*, *Citrobacter freundii*, *Stenotrophomonas* sp, *Flavobacterium* sp, *Proteus vulgaris*, *Pseudomonas* sp, *Acinetobacter* sp, *Klebsiella* sp, and *Proteus* sp are useful bacteria in the biodegradation of pesticides.
c. *Alcaligenes faecalis* JBW4 helps in organochlorine pesticide degradation.
d. In the degradation of chlordane and ϒ- chlordane, *Streptomyces* strains were considered as the best options.

2. Fungal Degradation

Fungi also form one of the classes of eukaryotic microbes that help in the biodegradation of POPs to maintain balance in the ecosystem. These are microscopic organism which can grow on various substrates. This group contains a large variety of organisms which include yeast, moulds, and filamentous fungi which produce different types of enzymes (*i.e.*, laccases, catalases, manganese peroxidase (MnP), Versatile peroxidase (VP), and lignin peroxidises (LiP) which help in the bioremediation of POPs. These are involved in the bioremediation of waste from various industrial wastewaters for hydrocarbons (*e.g.*, aliphatic and aromatic), chlorinated (*e.g.*, PCBs), phenolic compounds, pesticides, and environmentally persistent dyes (*e.g.*, Methylene blue). Few examples of fungi degradation are as follows:

a. *Irpexlacteus*and *Pleurotusostreatus* are responsible for the degradation and removal of PAHs.
b. *Mucor alternans, Fusarium oxysporum, Tricodermaviride* and *Phanerochae techrysosporium* had the ability for degradation of DDT.
c. *Tricodermaharzianum* degrades endosulfan pesticides.
d. *Phanerochae techrysosporium* is the best available option for a large range of the PAHs.

3. Algal Degradation

Algae come under the photosynthetic microbes as they can synthesise their foods on their own and produce some enzymes which help in the degradation of many organic pollutants. Algae are effective in the degradation of petroleum hydrocarbon into the less toxic forms, also the xenobiotics such as pesticides and PCBs. In photosynthetic organisms like algae, the degradation process completes in three phases. Some of the algae used for degradation are:

a. *Scenedesmus obliquus* GH2 for degradation of crude oil.
b. *Rhodococcus genus, Caepidiumantarticum, Desmarestia sp., Focus sp.* and *Ascophyllum nodosum* are effective in the degradation of hydrocarbons.
c. *Scenedesmus obliquus* and *Scenedesmus quadricauda* could remove broad-spectrum fungicides (*e.g.*, dimethomorph, pyrimethanil) and herbicides (*e.g.*, isoproturon).

OXIDATION PROCESS FOR WASTEWATER TREATMENT

Wastewater is mainly generated from the industries and the agriculture sector and household sewage, out of which industrial wastewater contains the highest number of organic pollutants. Wastewater contains many pollutants that are hazardous to human and environment both. For the removal of various POPs, convectional wastewater treatment is not a very good choice. Advanced oxidation process (AOPs) is highly an effective oxidation method that uses hydroxyl radicals or sulphate radicals for the remediation of organic pollutants from wastewater. In some AOPs hydroxyl radical uses ozone or UV irradiation as a catalyst. Some other AOPs such as electrochemical treatment, plasma, microwave, ultrasound-based methods are being used by many scientists. Hydroxyl radicals are highly reactive, short-lived species, immediately react/attack nearby organic pollutants and break them completely or in smaller molecules that are easily biodegradable.

RECENT ADVANCEMENTS

To increase efficiency and provide better outcomes, several research works are focused on designing and developing new technologies/methodologies for biodegradation and bioremediation with the help of genetically modified microbes and metagenomics. The studies are also conducted with the aid of microbial fuel cells, nanomaterials, biofilms, and constructed wetlands to degrade various POPs. As described earlier, microbes are used for the bioremediation of POPs. To further enhance the activity of microbes they are genetically modified for bioremediation purposes. In the present decade, nanomaterials are of great interest due to their unique features in removing pollutants from industrial effluents, thereby opening an area of development of hybrid nanoparticles for remediation. Few examples related to this are as follow:

a. *Bacillus megaterium* releases a wild type of cytochrome P450 enzyme which is not efficient enough for degradation of PAHs (*e.g.*, acenaphtylene, naphthalene, acenaphthene, 9-methlyanthracene, and fluorine). So, to increase its activity several mutations were done to get cytochrome P450 BM-3. The modified enzyme proved to show higher activity towards PAHs degradation in the environment.

b. $MnFe_2O_4$@PANI@Ag catalyst was designed for the decomposition of azo dyes (*e.g.*, Rhodamine B, Methylene Blue, and eosin Y).

c. The novel $FeNi_3/SiO_2/CuS$ (FNSCS) magnetic nanocomposite can be used for the degradation of tetracycline antibiotics.

d. *Aspergillus awamori* is used in COD at the pre-treatment stage.

CONCLUDING REMARKS

An ancient intruder (POPs) is composed of recalcitrant compounds. As discussed above, they can be accumulated in the adipose tissues, but recent research stated that they can also be accumulated in the proteins of the animals. So, the situations become even worse. The only way to find prevention from this situation is to synthesize their substitutes to reduce the risk. In addition to this, one should not only care to replace them with substitutes but also not develop other toxic chemical compounds in the future. Organic products made from natural products can be a healthy alternative and better substitute for POPs. Many research, regulatory efforts are made to fight against this misfortune, and we hope that one day these unintentional intruders will vanish from the very edge of this world.

LIST OF ABBREVIATIONS

PCDDs	Polychlorinated dibenzodioxins
PCDFs	Polychlorinated dibenzofurans
PBDDs	Polybrominated dibenzodioxins
PBDFs	Polybrominated dibenzofurans
PCBs	Polychlorinated biphenyls
PBDEs	Polybrominated diphenyl ethers
PCNs	Polychlorinated naphthalene's
PFAAs	Perfluoroalkyl acids
OCPs	Organochlorine pesticide
GC-ECD	Gas Chromatography Electron capture detector
GC-FID	Gas Chromatography Flame ionization detector
GC-NPD	Gas Chromatography Nitrogen Phosphorus detector
GC-TCD/MS	Gas Chromatography Thermal conductivity detector/Mass spectrometer
GC-MS/MS	Gas Chromatography - Mass spectrometer/ Mass spectrometer
LC-ESI/MS	Liquid Chromatography Electron spray ionization/ Mass spectrometer
HPLC-DAD	High-performance liquid Chromatography Diode array detector
ELISA	Enzyme-linked immunosorbent assay

CONSENT FOR PUBLICATION

Not Applicable.

CONFLICT OF INTEREST

The author confirms that this chapter contents have no conflict of interest.

ACKNOWLEDGEMENT

Declared none.

REFERENCES

[1]   Persistent Organic Pollutants. 2017.https://www.epa.gov/international-cooperation/persistent-organ-c-pollutants-global-issue-global-response

[2]   Jones KC, de Voogt P. Persistent organic pollutants (POPs): state of the science. Environ Pollut 1999; 100(1-3): 209-21.
[http://dx.doi.org/10.1016/S0269-7491(99)00098-6] [PMID: 15093119]

[3]　　WHO https://www.who.int/foodsafety/areas_work/chemical-risks/pops/en/

[4]　　Khairy MA, Luek JL, Dickhut R, Lohmann R. Levels, sources and chemical fate of persistent organic pollutants in the atmosphere and snow along the western Antarctic Peninsula. Environ Pollut 2016; 216: 304-13.
[http://dx.doi.org/10.1016/j.envpol.2016.05.092] [PMID: 27288629]

[5]　　Antignac JP, Main KM, Virtanen HE, *et al.* Country-specific chemical signatures of persistent organic pollutants (POPs) in breast milk of French, Danish and Finnish women. Environ Pollut 2016; 218: 728-38.
[http://dx.doi.org/10.1016/j.envpol.2016.07.069] [PMID: 27521295]

[6]　　Long M, Knudsen A-KS, Pedersen HS, Bonefeld-Jørgensen EC. Food intake and serum persistent organic pollutants in the Greenlandic pregnant women: The ACCEPT sub-study. Sci Total Environ 2015; 529: 198-212.
[http://dx.doi.org/10.1016/j.scitotenv.2015.05.022] [PMID: 26011616]

[7]　　Rylander C, Sandanger TM, Nøst TH, Breivik K, Lund E. Combining plasma measurements and mechanistic modeling to explore the effect of POPs on type 2 diabetes mellitus in Norwegian women. Environ Res 2015; 142: 365-73.
[http://dx.doi.org/10.1016/j.envres.2015.07.002] [PMID: 26208316]

[8]　　Ma J, Cao Z. Quantifying the perturbations of persistent organic pollutants induced by climate change. Environ Sci Technol 2010; 44(22): 8567-73.
[http://dx.doi.org/10.1021/es101771g] [PMID: 20923220]

[9]　　Sun LG, Yin XB, Pan CP, Wang YH. A 50-years record of dichloro-diphenyl-trichloroethanes and hexachlorocyclohexanes in lake sediments and penguin droppings on King George Island, Maritime Antarctic. J Environ Sci (China) 2005; 17(6): 899-905.
[PMID: 16465874]

[10]　Ma J. 201. Revolatilization of persistent organic pollutants in the Arctic induced by climate change. NatClimChange1 255-60.

[11]　Hallanger IG, Ruus A, Warner NA, *et al.* Differences between Arctic and Atlantic fjord systems on bioaccumulation of persistent organic pollutants in zooplankton from Svalbard. Sci Total Environ 2011; 409(14): 2783-95.
[http://dx.doi.org/10.1016/j.scitotenv.2011.03.015] [PMID: 21600630]

[12]　Chi KH, Lin CY, OuYang CF, *et al.* Evaluation of environmental fate and sinks of PCDD/Fs during specific extreme weather events in Taiwan. J Asian Earth Sci 2013; 77: 268-80.
[http://dx.doi.org/10.1016/j.jseaes.2013.04.006]

[13]　Zhou H, Wu H, Liao C, *et al.* Toxicology mechanism of the persistent organic pollutants (POPs) in fish through AhR pathway. Toxicol Mech Methods 2010; 20(6): 279-86.
[http://dx.doi.org/10.3109/15376516.2010.485227] [PMID: 20507254]

[14]　Mrema EJ, Rubino FM, Brambilla G, Moretto A, Tsatsakis AM, Colosio C. Persistent organochlorinated pesticides and mechanisms of their toxicity. Toxicology 2013; 307: 74-88.
[http://dx.doi.org/10.1016/j.tox.2012.11.015] [PMID: 23219589]

[15]　Damstra T, Page SW, Herrman JL, Meredith T. Persistent organic pollutants: Potential health effects? J Epidemiol Community Health 2002; 56(11): 824-5. b
[http://dx.doi.org/10.1136/jech.56.11.824] [PMID: 12388570]

[16]　Fontenele EG, Martins MR, Quidute AR, Montenegro RM Jr. Environmental contaminants and endocrine disruptors. Arq Bras Endocrinol Metabol 2010; 54(1): 6-16.
[http://dx.doi.org/10.1590/S0004-27302010000100003] [PMID: 20414542]

[17]　Crinnion WJ. Polychlorinated biphenyls: persistent pollutants with immunological, neurological, and endocrinological consequences. Altern Med Rev 2011; 16(1): 5-13.
[PMID: 21438643]

[18] Hertz-Picciotto I, Park HY, Dostal M, Kocan A, Trnovec T, Sram R. Prenatal exposures to persistent and non-persistent organic compounds and effects on immune system development. Basic Clin Pharmacol Toxicol 2008; 102(2): 146-54.
[http://dx.doi.org/10.1111/j.1742-7843.2007.00190.x] [PMID: 18226068]

[19] Vandelac L. [Endocrine disruption agents: environment, health, public policies, and the precautionary principle]. Bull Acad Natl Med 2000; 184(7): 1477-86.
[PMID: 11261252]

[20] Waissmann W. Health surveillance and endocrine disruptors. Cad Saude Publica 2002; 18(2): 511-7.
[http://dx.doi.org/10.1590/S0102-311X2002000200016] [PMID: 11923893]

[21] Wolff MS. Endocrine disruptors: challenges for environmental research in the 21st century. Ann N Y Acad Sci 2006; 1076: 228-38.
[http://dx.doi.org/10.1196/annals.1371.009] [PMID: 17119205]

[22] McLachlan JA, Arnold SF. Environmental Estrogens, American Scientist. accessible through http://www.amsci.org/amsci/articles/96articles/McLachla.html

[23] Jacobson SW, Fein GG, Jacobson JL, Schwartz PM, Dowler JK. The effect of intrauterine PCB exposure on visual recognition memory. Child Dev 1985; 56(4): 853-60.
[http://dx.doi.org/10.2307/1130097] [PMID: 3930167]

[24] Jacobson JL, Jacobson SW, Humphrey HEB. Effects of in utero exposure to polychlorinated biphenyls and related contaminants on cognitive functioning in young children. J Pediatr 1990; 116(1): 38-45. a
[http://dx.doi.org/10.1016/S0022-3476(05)81642-7] [PMID: 2104928]

[25] Jacobson JL, Jacobson SW, Humphrey HEB. Effects of exposure to PCBs and related compounds on growth and activity in children. Neurotoxicol Teratol 1990; 12(4): 319-26. b
[http://dx.doi.org/10.1016/0892-0362(90)90050-M] [PMID: 2118230]

[26] Schantz SL, Bowman RE. Learning in monkeys exposed perinatally to 2,3,7,8-tetrachlorodibenzo-p-dioxin (TCDD). Neurotoxicol Teratol 1989; 11(1): 13-9.
[http://dx.doi.org/10.1016/0892-0362(89)90080-9] [PMID: 2725437]

[27] Levin ED, Schantz SL, Bowman RE. Delayed spatial alternation deficits resulting from perinatal PCB exposure in monkeys. Arch Toxicol 1988; 62(4): 267-73.
[http://dx.doi.org/10.1007/BF00332486] [PMID: 3149182]

[28] Rice DC. Behavioral impairment produced by low-level postnatal PCB exposure in monkeys. Environ Res 1999; 80(2 Pt 2): S113-21.
[http://dx.doi.org/10.1006/enrs.1998.3917] [PMID: 10092425]

[29] Gao M, Wu NX. Male reproductive toxicity of polychlorinated biphenyls. Zhonghua Nan Ke Xue 2011; 17(5): 448-52.
[PMID: 21837958]

[30] Guo YL, Lambert GH, Hsu CC, Hsu MM. Yucheng: health effects of prenatal exposure to polychlorinated biphenyls and dibenzofurans. Int Arch Occup Environ Health 2004; 77(3): 153-8.
[http://dx.doi.org/10.1007/s00420-003-0487-9] [PMID: 14963712]

[31] Kumar S. Occupational exposure associated with reproductive dysfunction. J Occup Health 2004; 46(1): 1-19.
[http://dx.doi.org/10.1539/joh.46.1] [PMID: 14960825]

[32] Vizcaino E, Grimalt JO, Fernández-Somoano A, Tardon A. Transport of persistent organic pollutants across the human placenta. Environ Int 2014; 65: 107-15.
[http://dx.doi.org/10.1016/j.envint.2014.01.004] [PMID: 24486968]

[33] Vafeiadi M, Vrijheid M, Fthenou E, *et al.* Persistent organic pollutants exposure during pregnancy, maternal gestational weight gain, and birth outcomes in the mother-child cohort in Crete, Greece (RHEA study). Environ Int 2014; 64: 116-23.

[http://dx.doi.org/10.1016/j.envint.2013.12.015] [PMID: 24389008]

[34]   Damstra T. Potential effects of certain persistent organic pollutants and endocrine disrupting chemicals on the health of children. J Toxicol Clin Toxicol 2002; 40(4): 457-65.
[http://dx.doi.org/10.1081/CLT-120006748] [PMID: 12216998]

[35]   Dewan P, Jain V, Gupta P, Banerjee BD. Organochlorine pesticide residues in maternal blood, cord blood, placenta, and breastmilk and their relation to birth size. Chemosphere 2013; 90(5): 1704-10.
[http://dx.doi.org/10.1016/j.chemosphere.2012.09.083] [PMID: 23141556]

[36]   Kvist L, Giwercman A, Weihe P, *et al.* Exposure to persistent organic pollutants and sperm sex chromosome ratio in men from the Faroe Islands. Environ Int 2014; 73: 359-64.
[http://dx.doi.org/10.1016/j.envint.2014.09.001] [PMID: 25222300]

[37]   El-Shahawi MS, Hamza A, Bashammakh AS, Al-Saggaf WT. An overview on the accumulation, distribution, transformations, toxicity and analytical methods for the monitoring of persistent organic pollutants. Talanta 2010; 80(5): 1587-97.
[http://dx.doi.org/10.1016/j.talanta.2009.09.055] [PMID: 20152382]

[38]   Goldstein S. Beta-blocking drugs and coronary heart disease. Cardiovasc Drugs Ther 1997; 11 (Suppl. 1): 219-25.
[http://dx.doi.org/10.1023/A:1007711025487] [PMID: 9211014]

[39]   Chobanian AV, Bakris GL, Black HR, *et al.* Seventh report of the Joint National Committee on Prevention, Detection, Evaluation, and Treatment of High Blood Pressure. Hypertension 2003; 42(6): 1206-52.
[http://dx.doi.org/10.1161/01.HYP.0000107251.49515.c2] [PMID: 14656957]

[40]   Kearney PM, Whelton M, Reynolds K, Muntner P, Whelton PK, He J. Global burden of hypertension: analysis of worldwide data. Lancet 2005; 365(9455): 217-23.
[http://dx.doi.org/10.1016/S0140-6736(05)17741-1] [PMID: 15652604]

[41]   Ljunggren SA, Helmfrid I, Salihovic S, *et al.* Persistent organic pollutants distribution in lipoprotein fractions in relation to cardiovascular disease and cancer. Environ Int 2014; 65: 93-9.
[http://dx.doi.org/10.1016/j.envint.2013.12.017] [PMID: 24472825]

[42]   Lee DH, Lind PM, Jacobs DR Jr, Salihovic S, van Bavel B, Lind L. Polychlorinated biphenyls and organochlorine pesticides in plasma predict development of type 2 diabetes in the elderly: the prospective investigation of the vasculature in Uppsala Seniors (PIVUS) study. Diabetes Care 2011; 34(8): 1778-84. b
[http://dx.doi.org/10.2337/dc10-2116] [PMID: 21700918]

[43]   Valera B, Jørgensen ME, Jeppesen C, Bjerregaard P. Exposure to persistent organic pollutants and risk of hypertension among Inuit from Greenland. Environ Res 2013; 122: 65-73.
[http://dx.doi.org/10.1016/j.envres.2012.12.006] [PMID: 23375553]

[44]   Uemura H, Arisawa K, Hiyoshi M, *et al.* Prevalence of metabolic syndrome associated with body burden levels of dioxin and related compounds among Japan's general population. Environ Health Perspect 2009; 117(4): 568-73.
[http://dx.doi.org/10.1289/ehp.0800012] [PMID: 19440495]

[45]   Barouki R. Can xenobiotics accumulated in adipose tissue contribute to a carcinogenic risk? Ann Endocrinol (Paris) 2013; 74(2): 154-5.
[http://dx.doi.org/10.1016/j.ando.2013.03.022] [PMID: 23587352]

[46]   Yu HY, Guo Y, Zeng EY. Dietary intake of persistent organic pollutants and potential health risks *via* consumption of global aquatic products. Environ Toxicol Chem 2010; 29(10): 2135-42.
[http://dx.doi.org/10.1002/etc.315] [PMID: 20872674]

[47]   Fredslund SO, Bonefeld-Jørgensen EC. Breast cancer in the Arctic--changes over the past decades. Int J Circumpolar Health 2012; 71: 19155.
[http://dx.doi.org/10.3402/ijch.v71i0.19155] [PMID: 22901290]

[48]   Pestana D, Teixeira D, Faria A, Domingues V, Monteiro R, Calhau C. Effects of environmental organochlorine pesticides on human breast cancer: putative involvement on invasive cell ability. Environ Toxicol 2015; 30(2): 168-76.
[http://dx.doi.org/10.1002/tox.21882] [PMID: 23913582]

[49]   Ghisari M, Eiberg H, Long M, Bonefeld-Jørgensen EC. Polymorphisms in phase I and phase II genes and breast cancer risk and relations to persistent organic pollutant exposure: a case-control study in Inuit women. Environ Health 2014; 13(1): 19.
[http://dx.doi.org/10.1186/1476-069X-13-19] [PMID: 24629213]

[50]   Mathur V, Bhatnagar P, Sharma RG, Acharya V, Sexana R. Breast cancer incidence and exposure to pesticides among women originating from Jaipur. Environ Int 2002; 28(5): 331-6.
[http://dx.doi.org/10.1016/S0160-4120(02)00031-4] [PMID: 12437282]

[51]   Hardell L, Andersson SO, Carlberg M, *et al.* Adipose tissue concentrations of persistent organic pollutants and the risk of prostate cancer. J Occup Environ Med 2006; 48(7): 700-7.
[http://dx.doi.org/10.1097/01.jom.0000205989.46603.43] [PMID: 16832227]

[52]   Kitabchi AE, Umpierrez GE, Miles JM, Fisher JN. Hyperglycemic crises in adult patients with diabetes. Diabetes Care 2009; 32(7): 1335-43.
[http://dx.doi.org/10.2337/dc09-9032] [PMID: 19564476]

[53]   Diabetes Fact sheet N°312". 2014. WHO. October 2013. Retrieved 25 March..

[54]   Crinnion WJ, Crinnion ND. The role of persistent organic pollutants in the worldwide epidemic of type 2 diabetes mellitus and the possible connection to Farmed Atlantic Salmon (Salmo salar). Altern Med Rev 2011; 16(4): 301-13.
[PMID: 22214250]

[55]   De Tata V. Association of dioxin and other persistent organic pollutants (POPs) with diabetes: epidemiological evidence and new mechanisms of beta cell dysfunction. Int J Mol Sci 2014; 15(5): 7787-811.
[http://dx.doi.org/10.3390/ijms15057787] [PMID: 24802877]

[56]   Kuo CC, Moon K, Thayer KA, Navas-Acien A. Environmental chemicals and type 2 diabetes: an updated systematic review of the epidemiologic evidence. Curr Diab Rep 2013; 13(6): 831-49.
[http://dx.doi.org/10.1007/s11892-013-0432-6] [PMID: 24114039]

[57]   Lee DH, Porta M, Jacobs DR Jr, Vandenberg LN. Chlorinated persistent organic pollutants, obesity, and type 2 diabetes. Endocr Rev 2014; 35(4): 557-601.
[http://dx.doi.org/10.1210/er.2013-1084] [PMID: 24483949]

[58]   Airaksinen R, Rantakokko P, Eriksson JG, Blomstedt P, Kajantie E, Kiviranta H. Association between type 2 diabetes and exposure to persistent organic pollutants. Diabetes Care 2011; 34(9): 1972-9.
[http://dx.doi.org/10.2337/dc10-2303] [PMID: 21816981]

[59]   Lee DH, Lee IK, Song K, *et al.* A strong dose-response relation between serum concentrations of persistent organic pollutants and diabetes: results from the National Health and Examination Survey 1999-2002. Diabetes Care 2006; 29(7): 1638-44.
[http://dx.doi.org/10.2337/dc06-0543] [PMID: 16801591]

[60]   Lee DH, Lee IK, Jin SH, Steffes M, Jacobs DR Jr. Association between serum concentrations of persistent organic pollutants and insulin resistance among nondiabetic adults: results from the National Health and Nutrition Examination Survey 1999-2002. Diabetes Care 2007; 30(3): 622-8.
[http://dx.doi.org/10.2337/dc06-2190] [PMID: 17327331]

[61]   Lee DH, Jacobs DR Jr, Steffes M. Association of organochlorine pesticides with peripheral neuropathy in patients with diabetes or impaired fasting glucose. Diabetes 2008; 57(11): 3108-11.
[http://dx.doi.org/10.2337/db08-0668] [PMID: 18647952]

[62]   Lee DH, Steffes MW, Sjödin A, Jones RS, Needham LL, Jacobs DR Jr. Low dose of some persistent organic pollutants predicts type 2 diabetes: a nested case-control study. Environ Health Perspect 2010;

118(9): 1235-42.
[http://dx.doi.org/10.1289/ehp.0901480] [PMID: 20444671]

[63]   UNEP Stockholm Convention Protecting health and the environment from persistent organic pollutants. "The 12 initial POPs under the Stockholm Convention. http://chm.pops.int/TheConvention/ThePOPs/The12InitialPOPs/tabid/296/Default.aspx

[64]   UNEP Stockholm Convention Protecting health and the environment from persistent organic pollutants. "The 12 initial POPs under the Stockholm Convention. http://chm.pops.int/TheConvention/ThePOPs/TheNewPOPs/tabid/2511/Default.aspx

[65]   https://toxtown.nlm.nih.gov/chemicals-and-contaminants/persistent-organic

[66]   Furlani RPZ, Marcilio KM, Leme FM, Tfoun SAV. Analysis of pesticide residues in sugarcane juice using QuEChERS sample preparation and gas chromatography with electron capture detection. Food Chem 2011; 126: 1283-7.
[http://dx.doi.org/10.1016/j.foodchem.2010.11.074]

[67]   Tsiropoulos NG, Amvrazi EG. Determination of pesticide residues in honey by single-drop microextraction and gas chromatography. J AOAC Int 2011; 94(2): 634-44.
[http://dx.doi.org/10.1093/jaoac/94.2.634] [PMID: 21563700]

[68]   Farajzadeh MA, Afshar Mogaddam MR, Rezaee Aghdam S, Nouri N, Bamorrowat M. Application of elevated temperature-dispersive liquid-liquid microextraction for determination of organophosphorus pesticides residues in aqueous samples followed by gas chromatography-flame ionization detection. Food Chem 2016; 212: 198-204.
[http://dx.doi.org/10.1016/j.foodchem.2016.05.157] [PMID: 27374524]

[69]   Xiao Z, He M, Chen B, Hu B. Polydimethylsiloxane/metal-organic frameworks coated stir bar sorptive extraction coupled to gas chromatography-flame photometric detection for the determination of organophosphorus pesticides in environmental water samples. Talanta 2016; 156-157: 126-33.
[http://dx.doi.org/10.1016/j.talanta.2016.05.001] [PMID: 27260444]

[70]   Afshar Mogaddam MR, Farajzadeh MA. Acid-base reaction-based dispersive liquid-liquid microextraction method for extraction of three classes of pesticides from fruit juice samples. J Chromatogr A 2016; 1431: 8-16.
[http://dx.doi.org/10.1016/j.chroma.2015.12.059] [PMID: 26755415]

[71]   Mahpishanian S, Sereshti H, Baghdadi M. Superparamagnetic core-shells anchored onto graphene oxide grafted with phenylethyl amine as a nano-adsorbent for extraction and enrichment of organophosphorus pesticides from fruit, vegetable and water samples. J Chromatogr A 2015; 1406: 48-58.
[http://dx.doi.org/10.1016/j.chroma.2015.06.025] [PMID: 26129984]

[72]   Saraji M, Jafari MT, Mossaddegh M. Carbon nanotubes@silicon dioxide nanohybrids coating for solid-phase microextraction of organophosphorus pesticides followed by gas chromatography-corona discharge ion mobility spectrometric detection. J Chromatogr A 2016; 1429: 30-9.
[http://dx.doi.org/10.1016/j.chroma.2015.12.008] [PMID: 26709024]

[73]   Liu Y, Gao Z, Wu R, Wang Z, Chen X, Chan TD. Magnetic porous carbon derived from a bimetallic metal-organic framework for magnetic solid-phase extraction of organochlorine pesticides from drinking and environmental water samples. J Chromatogr A 2017; 1479: 55-61.
[http://dx.doi.org/10.1016/j.chroma.2016.12.014] [PMID: 27986286]

[74]   Machado I, Gérez N, Pistón M, Heinzen H, Cesio MV. Determination of pesticide residues in globe artichoke leaves and fruits by GC-MS and LC-MS/MS using the same QuEChERS procedure. Food Chem 2017; 227: 227-36.
[http://dx.doi.org/10.1016/j.foodchem.2017.01.025] [PMID: 28274427]

[75]   Shi Z, Hu J, Li Q, Zhang S, Liang Y, Zhang H. Graphene based solid phase extraction combined with ultra high performance liquid chromatography-tandem mass spectrometry for carbamate pesticides analysis in environmental water samples. J Chromatogr A 2014; 1355: 219-27.

[http://dx.doi.org/10.1016/j.chroma.2014.05.085] [PMID: 24973804]

[76]   Kujawski MW, Bargańska Ż, Marciniak K, Miedzianowska E, Kujawski JK, Ślebiodac M, *et al.* Determining pesticide contamination in honey by LC-ESIMS/MS - comparison of pesticide recoveries of two liquid-liquid extraction basedapproaches. Lebensm Wiss Technol 2014; 56: 517-23.
[http://dx.doi.org/10.1016/j.lwt.2013.11.024]

[77]   Yang X, Qiao K, Liu F, *et al.* Magnetic mixed hemimicelles dispersive solid-phase extraction based on ionic liquid-coated attapulgite/polyaniline-polypyrrole/Fe$_3$O$_4$ nanocomposites for determination of acaricides in fruit juice prior to high-performance liquid chromatography-diode array detection. Talanta 2017; 166: 93-100.
[http://dx.doi.org/10.1016/j.talanta.2017.01.051] [PMID: 28213265]

[78]   Qian G, Wang L, Wu Y, Zhang Q, Sun Q, Liu Y, *et al.* A monoclonal antibody- based sensitive enzyme-linked immunosorbent assay (ELISA) for the analysis of the organophosphorous pesticides chlorpyrifos-methyl in real samples. Food Chem 2009; 117: 364-70.
[http://dx.doi.org/10.1016/j.foodchem.2009.03.097]

[79]   Turner WE, Whitfield WE, Cash TP, *et al.* Organohalogen Compd 2011; 73: 611.

[80]   Zhang Z, Rhind SM. Optimized determination of polybrominated diphenyl ethers and polychlorinated biphenyls in sheep serum by solid-phase extraction-gas chromatography-mass spectrometry. Talanta 2011; 84(2): 487-93.
[http://dx.doi.org/10.1016/j.talanta.2011.01.042] [PMID: 21376977]

[81]   Focant J-F, De Pauw E, Chromatogr J. Chromatogr., B 776 (2002) 199.

[82]   Patterson DG Jr, Welch SM, Turner WE, Sjo¨din A, Focant J-F, Chromatogr J. A 1218 (2011) 3274.

[83]   Thomsen C, Liane VH, Becher G, Chromatogr J. B 846 (2007) 252.

[84]   Focant JF, Sjödin A, Turner WE, Patterson DG Jr. Measurement of selected polybrominated diphenyl ethers, polybrominated and polychlorinated biphenyls, and organochlorine pesticides in human serum and milk using comprehensive two-dimensional gas chromatography isotope dilution time-of-flight mass spectrometry. Anal Chem 2004; 76(21): 6313-20.
[http://dx.doi.org/10.1021/ac048959i] [PMID: 15516123]

[85]   Grimalt JO, Howsam M, Carrizo D, Otero R, de Marchi MRR, Vizcaino E. Integrated analysis of halogenated organic pollutants in sub-millilitre volumes of venous and umbilical cord blood sera. Anal Bioanal Chem 2010; 396(6): 2265-72.
[http://dx.doi.org/10.1007/s00216-010-3460-y] [PMID: 20135307]

[86]   Yeung LWY, Taniyasu S, Kannan K, *et al.* A 1216 (2009) 4950.

[87]   Keller JM, Calafat AM, Kato K, Ellefson ME, Reagen WK, Strynar M. S. OConnell, C.M. Butt, S.A. Mabury, J. Small, D.C.G. Muir, S.D. Leigh, M.M. Schantz. Anal Bioanal Chem 2010; 397: 439.
[http://dx.doi.org/10.1007/s00216-009-3222-x] [PMID: 19862506]

[88]   Sandau CD, Sjödin A, Davis MD, *et al.* Comprehensive solid-phase extraction method for persistent organic pollutants. Validation and application to the analysis of persistent chlorinated pesticides. Anal Chem 2003; 75(1): 71-7.
[http://dx.doi.org/10.1021/ac026121u] [PMID: 12530820]

[89]   Yusoff I, Alias Y, Yusof M, Ashraf MA. Assessment of pollutants migration at AmparTenang landfill site, Selangor, Malaysia. Sci Asia 2013; 39: 392-409.
[http://dx.doi.org/10.2306/scienceasia1513-1874.2013.39.392]

[90]   https://link.springer.com/article/10.1007/s11356-015-5225-9

CHAPTER 5

An Experimental and Simulation Study to Address Variabilities and Uncertainties in Risk Assessment of Lead and Cadmium Ingestion for a Vegetarian Diet

Ashish Yadav[1], Kaniska Biswas[2], Mukesh Sharma[3,*], Arunima Khare[3] and **Pavan K. Nagar[3]**

[1] *Department of Chemical Engineering, Indian Institute of Technology Kanpur, Kanpur, India*

[2] *Design Programme, Indian Institute of Technology Kanpur, Kanpur, India*

[3] *Department of Civil Engineering, Center for Environmental Science and Engineering, Indian Institute of Technology Kanpur, Kanpur, India*

Abstract: The risk of adverse health effects of heavy metals, lead (Pb) and cadmium (Cd), was characterized by considering dietary intake of food items and resulting levels of biomarkers, blood Pb levels (PbB), and urinary Cd levels (CdU). Specifically, 35 food items (cereals, pulses, vegetables, and fruits), used in a vegetarian diet in India, were considered. Samples of food items were taken in the winter and autumn seasons and were analysed for Pb and Cd. The observed concentrations were translated into probability density functions (PDF) and Monte Carlo simulation was used to generate levels of chronic daily intake (CDI) that accounted for variability in (i) body weight, (ii) concentration in food and (iii) amount of food intake. The CDI levels were translated into equivalent PDF and the probabilities of exceedance of WHO-suggested provisional tolerable weekly intake of Pb and Cd were estimated. The probability of exceedance of the WHO tolerable limit was 5.55×10^{-3} for Pb and 7.36×10^{-4} for Cd. Further, CDIs were translated into PbB levels using a physiologically based pharmacokinetic model. The estimated health risk from (i) ingestion of Pb (*i.e.*, probability of exceedance of safe PbB level of 10 µg/dL) was 9.24×10^{-3} and (ii) ingestion of Cd (*i.e.*, probability of exceedance of 5 µg/g creatinine) was 4.21×10^{-5}, suggesting that Pb in the environment still poses a substantial risk despite its phasing out from gasoline.

Keywords: Biomarker, Cadmium, Exposure, Lead, Risk.

* **Corresponding author Mukesh Sharma:** Department of Civil Engineering, Indian Institute of Technology Kanpur, Kanpur, India; E-mail:mukesh@iitk.ac.in

Tahmeena Khan, Abdul Rahman Khan, Saman Raza, Iqbal Azad and Alfred J. Lawrence (Eds.)

INTRODUCTION

Lead (Pb) and cadmium (Cd) metals and their compounds are used in industrial and domestic products. Pb is used in Pb-based paints, Pb-containing pipes in water supply systems, battery recycling, plastics, ceramics, *etc* [1]. Cd is used in nickel-Cd batteries, as a pigment, for corrosion-resistant plating on steel, and to stabilize plastic [2]. Both the heavy metals find their way in the environment and show ubiquitous presence, even far from their emitting sources (mining, smelters, refining, incineration, and waste disposal). These metals are persistent and toxic and have no useful biological function in the human body. Even small quantities of these metals may pose a health risk to humans and the environment.

Pb is listed as one of the hazardous heavy metals by the Agency for Toxic Substances and Disease Registry (ATSDR) [3]. The main source of Pb in the air was through vehicular emissions, but after the introduction of unleaded gasoline in 2000, the air Pb levels have significantly dropped in Indian cities. For example, in Kanpur (Latitude 26.4670° North and Longitude 80.3500° East), the Pb levels during the pre-unleaded period were 2.0-2.75 $\mu g/m^3$, which have dropped to about 0.70 $\mu g/m^3$ [3 - 6]. However, Pb is still found in various environmental media (air, water, and soil) due to its stable and persistent nature. Once Pb enters the food chain, it continues to be found at all trophic levels and poses a significant health risk. In Japanese cities, it continues to be present in various food items even after 20 years of the development of unleaded gasoline [7].

The biomarker useful in estimating Pb body burden is blood lead (PbB) level. In response to increasing epidemiological evidence, the Centres for Disease Control and Prevention (CDC) gradually lowered the acceptable PbB level in the US from 60 $\mu g/dL$ before 1975 to 10 $\mu g/dL$, in 1991. As per the CDC, there is no safe level of PbB below which there may not be any adverse effects [8]. The World Health Organization (WHO) has also proposed an acceptable PbB level of 10 $\mu g/dL$, which has been adopted by several countries for developing air quality standards [9]. There is poor intellectual development in children at PbB levels even below 10 $\mu g/dL$ [10, 11]. Prenatal Pb exposure is reported to have a more lasting impact on child development than postnatal exposure [11]. The study by Schnaas *et al.* [11] also suggests that PbB levels in expectant mothers should be well below 10 $\mu g/dL$.

Physiologically based pharmacokinetic (PBPK) models are frequently used to assess the risk of exposure to metals and other toxins. The PBPK models describe the relationships among critical biological processes using mathematical mass balance equations that account for the disposition and uptake of chemical substances [12]. Estimates of the chemical-specific physicochemical, physiological, and biological parameters are required to apply the PBPK model for any substance. A PBPK model for Pb has been developed to estimate the PbB (the biomarker for Pb exposure) concentration and assess the health risk due to Pb [5, 13].

Cd and its compounds have been established as toxins having adverse health effects by the US Department of Health and Human Services and the International Agency for Research on Cancer [14, 15]. Cd exists as vapours and/or particles in the air and it may be deposited either by wet or dry deposition onto soil and water surfaces. Cd present in the soil is largely immobile since it binds strongly to the organic matter and enters the food chain. In water, the soluble form migrates, whereas the insoluble part gets accumulated in sediment [14]. The biological indicators of Cd exposure are Cd levels in the blood, urine, kidney, liver, hair, faeces, and other tissues. Cd level in urine (CdU) reflects the total body burden of Cd [16]. The biological half-life of Cd is over 13 years [17].

To assess the risk due to heavy metal exposure, it is important to account for variability in metal concentration in food, water, and air. Information on the average level is of little use in risk assessment. A probabilistic assessment of exposure to heavy metal through ingestion and inhalation, exceeding the safe exposure level (*e.g.*, provisional tolerable weekly intake (PTWI)), by Herman and Younes [18] indicates the underlying risk [19]. The estimation of the probability of interest depends on the tail behaviour of the statistical distribution of exposure, more precisely on the extreme exposures [20]. Under the complexity of the distribution of different variables, exposure can be assessed through Monte Carlo simulation, which accounts for several variables at a time [21]. The variables include body weight, food consumption quantity and metal concentration in various food items. As the numbers of food items are many, there is a need to group them; notably, those grown under and above the ground. The present research objective is to characterize the risk of the adverse health effect of Pb and Cd for the vegetarian population of Kanpur city. The specific objectives of this research were to (i) measure concentrations of Cd and Pb in food products (fruits, vegetables, cereals, pulses, milk) and water, (ii) estimate product-wise food consumption, and (iii) develop statistical distributions of intake of Pb and Cd and characterize the risk through their biomarkers, PbB and CdU, by employing the Monte Carlo simulation.

STUDY AREA, MATERIALS, AND METHOD

The study area for the present research was Kanpur city (population – over 37,00,000; area - 3,029 km^2) (Fig. **1**), which is characterized by many industries (leather, paint, soaps, and detergents, *etc.*), including a 220 MW coal-based thermal power plant. The study has three parts: (i) collection of samples of food items and water, (ii) laboratory analysis of the samples for Pb and Cd levels and (iii) modelling, simulations, and risk characterization.

SAMPLE COLLECTION

Food and Water Samples

The sampling of the various food items was done from four major food markets: (i) Kalyanpur (KYP), (ii) Nawabganj (NBJ), (iii) Vijaynagar (VJN) and (iv) Ramadevi (RMD) (Fig. **1**); the samples were collected twice, one in each of the two seasons (autumn and winter). The samples were collected according to the market basket method [21]. This method simulates how consumers purchase their food items.

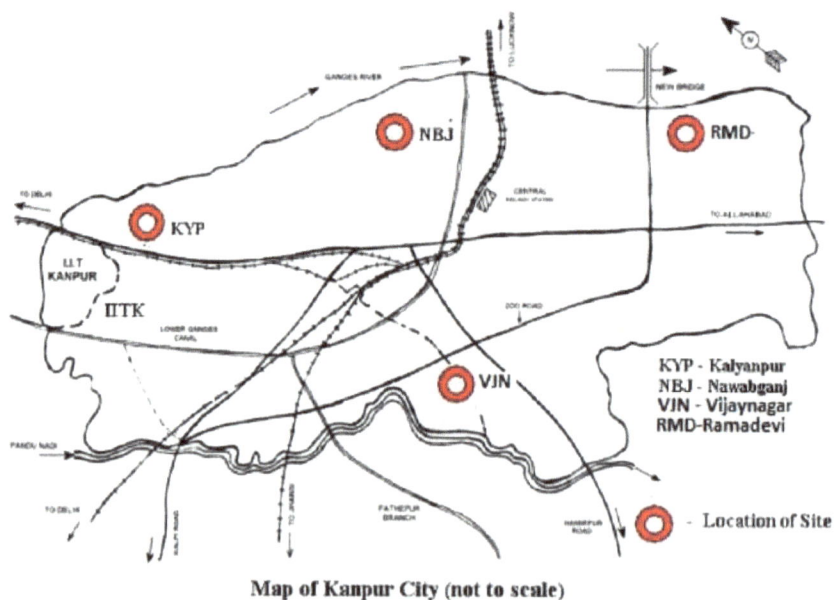

Map of Kanpur City (not to scale)

Fig. (1). Kanpur City: Sampling locations for collection of food items.

The food items were distributed in seven different categories: (i) leafy vegetables, (ii) non-leafy vegetables (underground), (iii) non-leafy vegetables (above ground), (iv) cereals, (v) pulses, (vi) fruits and (vii) milk (Tables **1** and **2**).

The metal uptake by vegetables can vary significantly for those grown above and underneath the ground. The adopted probability distinction accounts for the variability in vegetables grown above and underneath the ground. The water samples were collected in clean plastic torsion bottles (Table **2**).

Table 1. Food items under different categories.

Food Category	Food Items
Leafy vegetables	Spinach, Coriander, Fenugreek
Non-leafy underground Vegetables	Potato, Carrot, Onion, Beetroot, Radish
Non-leafy above-ground vegetables	Pumpkin, Brinjal, Beans, Tomato, Chilli, Pea, Ladyfinger, Cabbage, Gourd, Cauliflower
Cereals	Rice, Wheat
Fruits	Apple, Papaya, Grapes, Orange, Banana
Pulses	Chana, Rajma, Arhar, Masoor, Moong, Urad (Black), Urad (green)
Milk	Buffalo Milk, Cow Milk, Dairy Milk

Table 2. Number of samples from different locations.

Food Category	Site				TOTAL
	KYP	NBJ	VJN	RMD	
Leafy vegetables	9	6	6	6	27
Non-leafy underground vegetables	10	10	10	10	40
Non-leafy above ground vegetables	16	16	16	18	66
Cereals	4	4	4	4	16
Fruits	5	5	5	5	20
Pulses	7	7	7	7	28
Municipal water	1	3	3	1	8
Ground water	4	2	3	4	13

URINE SAMPLES

Prior approval of the methodology of urine collection was sought from the Ethics Committee of the Indian Institute of Technology, Kanpur, which required written consent from subjects who volunteered to provide their urine samples. The urine samples were collected from 40 subjects (male non-smokers only (18 - 66 years of age) across Kanpur city and their body weight and age were recorded. Clean wide mouth polyethylene bottles were used to collect the spot samples. The first discarded samples were collected and preserved by adding 1% 6 M hydrochloric acid at 4°C [22].

SAMPLE ANALYSIS

Sample Processing

The processing of the various food items was done as described by IEPHM 1991 [23]. Stainless steel knife was used for removing the peels of vegetables and fruits and then the edible portion was thoroughly washed thrice with distilled water. Each of the collected samples was then cut into fine pieces and placed in Petri-dishes. The moisture content was determined by taking initial and final weights (after drying at 105 °C for 12 hours). The dried samples were then turned into a fine powder with the help of a mixer grinder. Similarly, the processing of pulses and cereals was also carried out. Water samples were directly analysed after filtering through Millipore 0.22-micron filter paper.

INSTRUMENTATION AND ANALYSIS

The microwave-assisted extraction of Pb and Cd was carried out following the USEPA Method 3052 [24]. 0.5 grams of dry powdered sample was digested with 9 ml of concentrated nitric acid for 15 minutes using a laboratory microwave (Multiwave 3000 Microwave Reaction System, Anton Paar GmbH, Austria). The sample was heated to 180 ± 5 °C in less than 5.5 minutes and then the temperature was kept constant at 180 ± 5 °C for the next 9.5 minutes to achieve complete digestion. The contents could cool and then were filtered through Millipore 0.22-micron filter paper. The extracted samples were then diluted with triple distilled de-ionized water to make up to 500 mL and stored in clean plastic torsion bottles. Along with each set of samples for digestion, one blank consisting of 9 mL of concentrated nitric acid and 0.5 mL of triple distilled de-ionized water was also digested. This blank was used as the method blank for analysis.

All the samples, including water, were analysed on Inductively Coupled Plasma-Mass Spectrometer (ICP-MS, Thermo Scientific XSeries 2, USA). The instrument was calibrated before analyzing the samples. A mixed standard of metals having a concentration of 100 ppm was used as a stock solution. The working standard of 100 ppb was prepared from this stock solution and was used to make further dilutions of 1 ppb, 2 ppb, 4 ppb, and 8 ppb. A calibration curve was prepared from these working standards.

The recovery of the extraction procedure was calculated using the post-digestion spike method. In this method, an aliquot of the test sample was added with a known amount of standard such that the heavy metal concentration of the resulting spiked sample becomes 2 to 5 times of what could be in the original test sample. The recoveries of the heavy metals were calculated at two different spike levels (5 and 8 ppb). Spiked samples were analysed and average recoveries obtained for Pb, Cd were 97% and 98%, respectively.

URINE ANALYSIS

Before extraction, urine samples could come to room temperature. After thoroughly mixing the sample, 3 mL of urine was taken in an inert polymeric microwave vessel and was digested in a microwave-assisted digestion system. The digestion procedure and temperature profile were like the ones used for the digestion of food samples. The ICP-MS was calibrated, and urine samples were analysed for heavy metal concentration.

RISK ASSESSMENT METHODOLOGY

For risk assessment, extreme pollutant concentrations are of paramount interest to estimate the probability of exceeding the safe doses and resulting levels of biomarker. Since measured concentrations are limited, one is not sure if extreme concentrations would always be considered. Thus, it becomes necessary to use a simulation model that can generate concentrations in all ranges (in a probabilistic sense). To accomplish the simulation, the observed data on variables (*e.g.*, metal concentration in food items, body weight, amount of food consumed) are used to estimate probabilistic distribution functions (PDF) and parameters of distribution (*e.g.*, mean and variance) of the variables.

The CDI for a pollutant (Pb or Cd in this study) i can be estimated as a summation of CDIs from j number of sources (*e.g.*, food items, water) as stated in equation **1**:

$$\text{CDI}_i(\mu g/kg\text{-}day) = \sum_j \frac{\left(I_{ij} \times C_{ij}\right)}{BW} \qquad (1)$$

Where I is dietary intake (kg/d), C is the concentration ($\mu g/kg$) and BW is body weight (kg).

The above formulation accounts for exposure frequency (365 d/y), exposure duration (ED, y), and averaging time (ED×365 d); for more details, one may refer to Asante-Duah [25].

One should consider the uncertainties in the following variables: (a) concentration of metal in all food items (C_{ij}), (b) consumption quantity of food items (I_{ij}), and (c) body weight (BW) for generating CDI using equation [1]. The random realizations of variables using the Monte Carlo simulation tool [26] can be used to generate the distribution of CDI. The risk is characterized by estimating the area exceeding the WHO limit (described later) from the generated PDF of CDI.

A PBPK model has been used in this study for estimating PbB levels. The model solves set of Pb mass balance differential equations having inputs of ambient air Pb concentration (0.70 µg/m^3), body weight, pharmacokinetic parameters, and the CDIs [5]. The validation of the model was done with the experimental data reported by Rabinowitz *et al.* [27] and Azar *et al.* [28]. The model has also been validated in the previous studies reported for the Kanpur city [5, 13]. Random 10,000 CDI levels were generated and translated to PbB levels using the PBPK model. The modelled PbB levels were transformed into PDF and risk was estimated as the probability of exceedance of level of PbB being greater than 10 µg/dL [8].

RESULTS AND DISCUSSION

Concentration of Pb and Cd in Various Food Items

The concentrations of Pb and Cd in various food items in Kanpur city and their comparison with other studies are presented in Table **3**. The Pb and Cd levels in vegetables showed a large variation among cities. The reason can be attributed to the fact that these are grown locally in varying environmental conditions. However, cereals and pulses did not show large variations because they are grown centrally and distributed to distant places. Fruits showed a minimal concentration of Pb and Cd. This may be because fruits come from faraway places, generally from rural areas.

In vegetables, the largest concentration of Pb and Cd was found in leafy vegetables [29 - 31]. The reason may be attributed to the rapid growth of leafy vegetables, which may absorb more of the nutrients and pollutants in a short time. Non-leafy underground vegetables have more Pb and Cd content than overground grown vegetables [32]. This variation may be because the underground vegetables have a larger uptake from the soil, which is not the case for above-ground vegetables.

Table **3** shows a comparison of the concentration of Pb in various food items for different areas of Kanpur city. The concentration of Pb in various food items has decreased over a period of time, but present levels may still pose a risk to human effect. However, in the case of non-leafy vegetables, there is a considerable decrease in the concentration of Pb compared to other food items.

To calculate CDI, the diet consumption data were taken from Sharma and Reddy [13] except for non-leafy vegetables. In the present study, underground and above-ground non-leafy vegetables were considered separately. Based on the consumption pattern of one month observed in the mess of student hostel-7, Indian Institute of Technology Kanpur, the consumption amount of underground non-leafy vegetables was 78±11.89 g/day and for above-ground non-leafy vegetables, it was 40±6.10 g/day.

In characterizing the risk, in addition to the variability in food consumption, the variation in body weight was also considered. The body weight data were found to be normally distributed (μ=68.85 kg, σ^2= 24.56 kg^2).

Table 3. Comparison of mean Pb and Cd concentration in food items with other cities (µg/kg unless specified otherwise)

Food Items		Present Study Kanpur		Urban Area Kanpur [5]	Rural Area Kanpur [13]	Bombay [33]		China [34]	
		Pb	**Cd**	**Pb**	**Pb**	**Pb**	**Cd**	**Pb**	**Cd**
Cereals		110.1 (23-210) N=16	17.4 (4-35) N=16	119.9±82.1 (28–223) N=8	106.4±80.1 (25-207) N=65	18.2 N=15	7.1 N=15	56.4 (4–616) N=59	13 (2-100) N=59
Pulses		225.5 (60-371) N=28	13.9 (4-25) N=28	283.3±118.4 (65–415) N=21	220.8±116.5 (43-405) N=48	253.3 N=13	7.9 N=13	33.0 (4–143) N=34	88.2 (7-381) N=34
Leafy Vegetables		227.4 (62-553) N=27	29.0 (0-65) N=27	325.6±74.1 (181– 541) N=32	317.7±61.8 (191- 437) N=15	100.4 N=11	14.9 N=11	-	-
Non-leafy vegetables	Above ground	15.4 (0-70) N=66	6.2 (0-123) N=66	121.9±58.3 (24–279) N=114	101.9±52.8 (36-243) N=66	4.1 N=32	3.2 N=32	-	-
	Under ground	36.0 (0-203) N=40	26.9 (0-332) N=40						
Fruits		4.2 (0-18) N=20	1.4 (0-2) N=20	7.3±6.1 (2–17) N=10	5.6±1.8 (1-17) N=26	7.1 N=7	0.1 N=7	-	-
Milk (µg/L)		1.7 (0-3) N=10	0.8 (0-1) N=10	4.1±2.8 (0–7) N=8	0.5±0.2 (ND-0.65) N=4	1.6 N=4	0.1 N=4	-	-
Water (µg/L)		1.7 (0-13) N=21	0.05 (0-0.39) N=21	8.4±3.9 (4–16) N=11	3.9±0.9 (3-5) N=6	1.2 N=13	0.01 N=13	-	-
The range in parenthesis shows the minimum and maximum level and N shows the number of samples.									

Risk Characterization

Dietary Intake (Pb and Cd)

The probability plots of Pb and Cd contents in various food items were obtained by performing the Kolmogorov-Smirnov test [35]. The PDF for Pb and Cd levels were as follows: (i) for Pb; cereals-lognormal, pulses-normal, vegetables-lognormal, water-lognormal, fruits-lognormal, milk-normal and (ii) for Cd; cereals-lognormal, pulses-normal, vegetables-lognormal, water-lognormal, fruits-normal, and milk-normal (Fig. **2**).

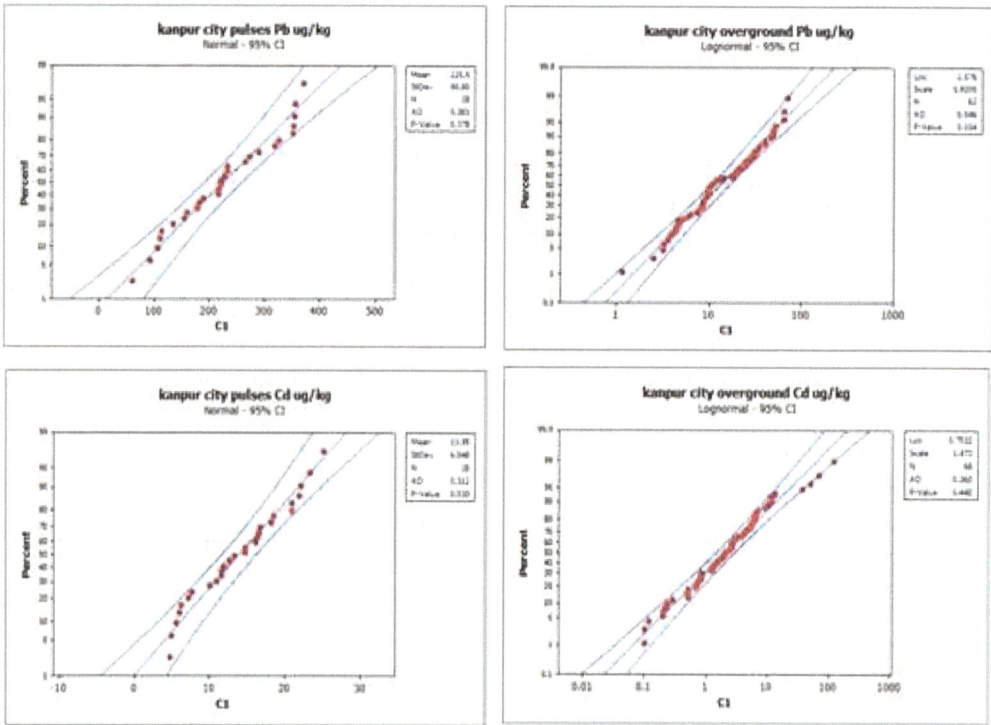

Fig. (2). Probability plots using the Kolmogorov-Smirnov test for some food items.

Based on the probability plots, 10000 random Pb and Cd levels were generated using Monte Carlo simulation. Similarly, 1000 random body weights were generated using the data of body weight, which was found to be normally distributed. The randomly generated Pb and Cd contents and body weights were translated into CDI using equation [1].

The mean contribution to CDI from different food items is as follows: cereals 62%, pulses 15%, vegetables 18%, water 4%, and fruits/milk <1%. More than 75% of the intake can be attributed to cereals and pulses due to their large consumption and these may be grown in areas where the soil is contaminated. There is a need to identify the contaminated areas and either decontaminate the soil or move the cultivation to the fields which are not contaminated.

FAO/WHO [36] has recommended the following (i) PTWI for Pb: 25 µg/kg (body weight)/week *i.e.*, 3.571 µg/kg (body weight)/day and (ii) PTMI for Cd: 25 µg/kg (body weight)/week *i.e.*, 0.833 µg/kg (body weight)/day. Although the estimated mean daily intake of Pb and Cd was less than the tolerance limit of WHO, this does not mean that the risk is zero or insignificant. The proper approach is to examine the probability distribution of generated CDI and assess the probability of exceedance of PTWI/PTMI. Figs. (**3** and **4**) show PDF of CDI for Pb and Cd. The probability of exceeding the WHO limit is 5.55×10^{-3} for Pb and 7.36×10^{-4} for Cd. These probabilities suggest impending risk due to exposure of Pb and Cd. It may be noted that the intake of Cd through the air route (less than 4% of total intake) was not considered due to low concentration (0.02 ± 0.022 µg/m^3) [37].

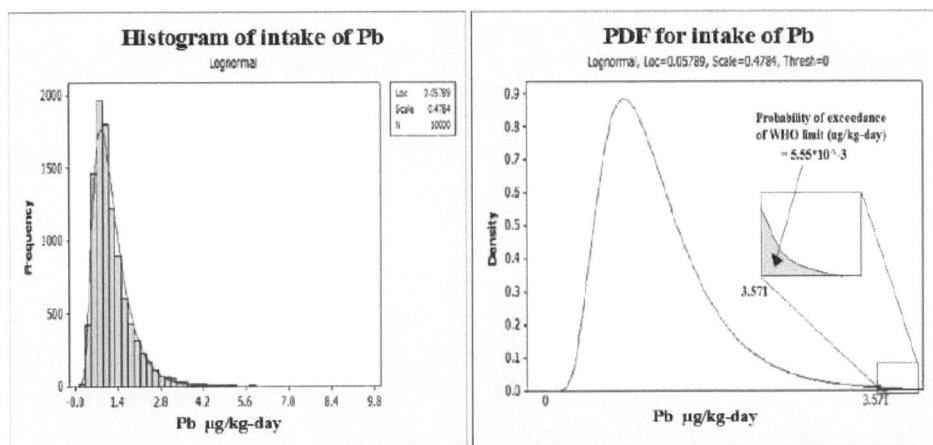

Fig. (3). Histogram and PDF of dietary intake of Pb and probability of exceedance of WHO limit.

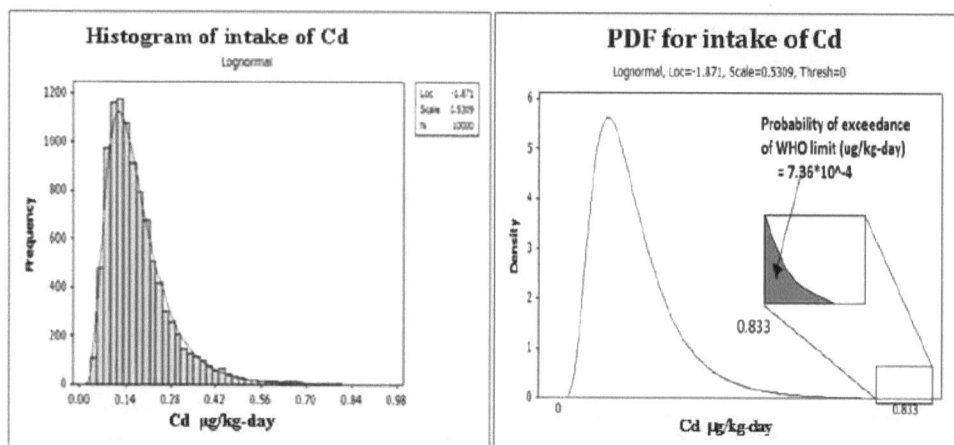

Fig. (4). Histogram and PDF of dietary intake of Cd and probability of exceedance of WHO limit.

PbB-based Risk Assessment

The risk of Pb is characterized based on the PBPK model, which estimated the PbB levels. The histogram and PDF of PbB levels are shown in Fig. (5). The estimated risk (*i.e.*, probability of PbB being greater than 10 µg/dL) is 9.24×10^{-3}. It is noteworthy that suggestive risks based on exceeding safe levels of CDI (Fig. 3) and from PbB simulations are comparable. It is a matter of concern that even after about 15 years of unleaded gasoline, there is still a significant risk of Pb.

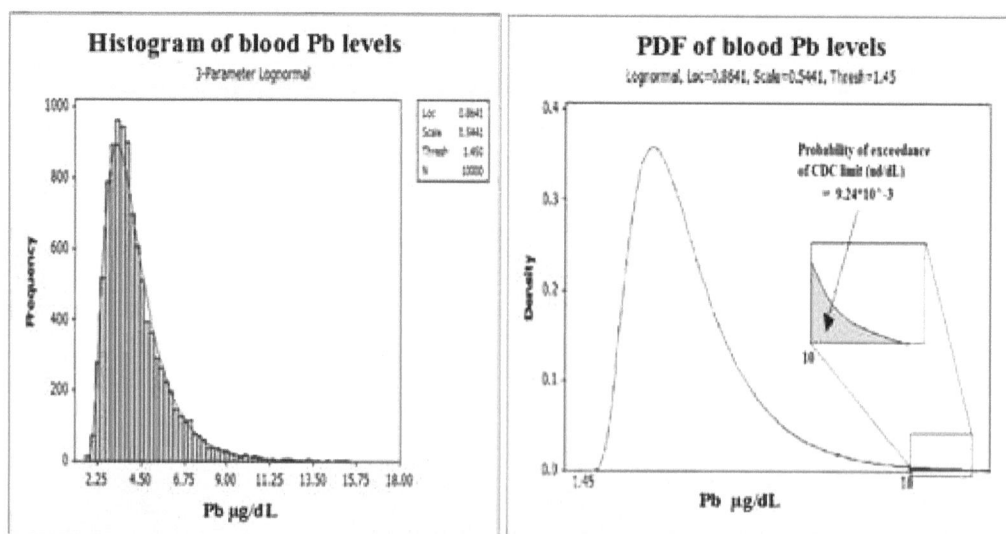

Fig. (5). Histogram and PDF of PbB levels and the probability of exceedance of WHO limit.

CdU-based Risk Assessment

The WHO [38] recommends that control measures should be taken if the level of CdU exceeds 5 µg Cd/g creatinine. The rate of creatinine excretion (Crur, g creatinine/day) was estimated from the relationship between lean body mass (LBM, kg) and Crur [39]:

LBM=27.2Crur + 8.58

Lean body mass for adult males was estimated as follows [40]:

LBM=0.88*BW (the mean BW of subjects was 68.85 kg)

Crur = 1.91 g Cr/day

For an average daily urine discharge of 1.5 L/day,

CdU (µg/g Cr) = CdU (µg/L) x (1.5 L/day)/(1.91 g Cr/day)

Table 4 shows the CdU levels in the present study, along with levels reported in other studies. CdU levels for the population of Kanpur city were log-normally distributed (Fig. 6). The estimated risk (*i.e.*, probability of CdU being greater than the WHO limit of 5 µg/g Cr) is 4.21×10^{-5}. The CdU based risk is one order smaller than the risk-based on CDI exceeding PTMI (Fig. 4); CdU based risk was from limited data of 40 subjects and did not consider uncertainties in variables.

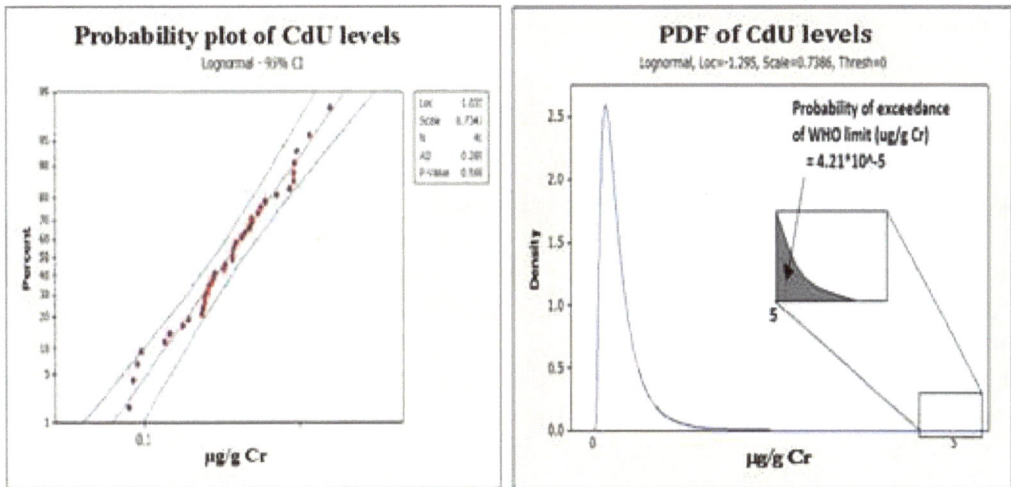

Fig. (6). Probability plot of CdU levels and the probability of exceedance of WHO limit.

Table 4. CdU levels in the present study and in other countries.

Sex	CdU (μg/g Cr[a])	Survey Location	Reference
Men	0.28±1.98	India	Present study
X uk[b]	0.40[c]	China	Nordberg *et al.* [41]
Men	0.59[d]	Belgium	Staessen *et al.* [42]
Men	0.90[c]	Belgium	Sartor *et al.* [43]
Women	0.49	Belgium	Staessen *et al.* [42]
Women	0.80[c]	Belgium	Sartor *et al.* [43]
X uk[b]	0.50-0.87	The Netherlands	Van Sittert *et al.* [44]
X uk[b]	0.59	Europe	Taylor *et al.* [45]
X uk[b]	0.66	Belgium	Staessen *et al.* [46]
X uk[b]	0.69	Belgium	Roels *et al.* [47]
X uk[b]	0.86	Germany	Jung *et al.* [48]
Women	1.47	Bangkok	Zhang *et al.* [49]
Women	1.44	Southeast Asia	Higashikawa *et al.* [50]
Women	1.29	China	Higashikawa *et al.* [50]
Women	1.65	Japan	Higashikawa *et al.* [50]
Women	1.70	Korea	Higashikawa *et al.* [50]

CONCLUDING REMARKS

The risk due to Pb and Cd in Kanpur city was characterized through dietary intake, based on WHO suggested PTWI/PTMI and levels of biomarkers (blood Pb level (PbB) and urinary Cd level (CdU)). The Pb and Cd levels in various food items (cereals, pulses, leafy vegetables, non-leafy underground vegetables, non-leafy above-ground vegetables, fruits, milk) and water showed variability, which was addressed by employing Monte Carlo simulation in the estimation of CDI. Due to its stable and persistent nature, Pb poses significant health risks {5.55×10^{-3} (probability of exceedance of PTWI) and 9.24×10^{-3} (probability of level of PbB being greater than 10 μg/dL)} even after the introduction of unleaded gasoline in the year 2000. The risk from Cd was also significant {7.36×10^{-4} (probability of exceedance of PTMI) and 4.21×10^{-5} (probability of level of CdU being greater than 5 μg/g Cr). The major intake (> 75%) of Pb and Cd is from cereals and pulses, suggesting contamination of fields where cereals and pulses are being grown, and measures are required for alternate fields and/or decontamination of soil.

CONSENT FOR PUBLICATION

Not Applicable.

CONFLICT OF INTEREST

The author confirms that this chapter contents have no conflict of interest.

ACKNOWLEDGEMENT

Declared none.

REFERENCES

[1]   Flora G, Gupta D, Tiwari A. Toxicity of lead: A review with recent updates. Interdiscip Toxicol 2012; 5(2): 47-58.
       [http://dx.doi.org/10.2478/v10102-012-0009-2] [PMID: 23118587]

[2]   Morrow H. Cadmium and cadmium alloys. Kirk-Othmer Encyclopedia of Chemical Technology 2010; pp. 1-36.

[3]   ATSDR (US Agency for Toxic Substance and Disease Registry). http://www.atsdr.cdc.gov /clist.html2002.

[4]   Malviya R, Wagela DK. Studies on lead concentration in ambient air, roadside dust and its influence on the healthy traffic police personnel at Indore city. Pollut Res 2001; 20: 635-8.

[5]   Sharma M, Maheshwari M, Morisawa S. Dietary and inhalation intake of lead and estimation of blood lead levels in adults and children in Kanpur, India. Risk Anal 2005; 25(6): 1573-88.
       [http://dx.doi.org/10.1111/j.1539-6924.2005.00683.x] [PMID: 16506983]

[6]   Behera SN, Sharma M. Reconstructing primary and secondary components of PM2.5 composition for an urban atmosphere. Aerosol Sci Technol 2010; 44: 1-11.
       [http://dx.doi.org/10.1080/02786826.2010.504245]

[7]   Moriwasa S, Hidaka H, Yoneda M. Historical review of health risk aspects with the leaded gasoline regulation in Japan, based on PBPK model. J Global Environ Eng 2001; 7: 63-78.

[8]   CDC. (Center of Disease Control) Preventing lead poisoning in young children. US Department of Health and Human Services 1991.

[9]   WHO (World Health Organization). Air quality guidelines for Europe. Regional Office for Europe. WHO Regional Publications, European Series, No 91, Copenhagen, Denmark..

[10]   Canfield RL, Henderson CR Jr, Cory-Slechta DA, Cox C, Jusko TA, Lanphear BP. Intellectual impairment in children with blood lead concentrations below 10 microg per deciliter. N Engl J Med 2003; 348(16): 1517-26.
       [http://dx.doi.org/10.1056/NEJMoa022848] [PMID: 12700371]

[11]   Schnaas L, Rothenberg SJ, Flores MF, *et al.* Reduced intellectual development in children with prenatal lead exposure. Environ Health Perspect 2006; 114(5): 791-7.
       [http://dx.doi.org/10.1289/ehp.8552] [PMID: 16675439]

[12]   ATSDR (US Agency for Toxic Substance and Disease Registry). 1999.

[13]   Sharma M, Reddy SU. Estimation of lead elimination rates for liver and kidney using a physiologically-based pharmacokinetic model for improved risk analysis. Hum Ecol Risk Assess 2012; 18: 627-48.
       [http://dx.doi.org/10.1080/10807039.2012.672899]

[14]   ATSDR (US Agency for Toxic Substance and Disease Registry). 2012.

[15]   IARC (International Agency for Research on Cancer). Cadmium and certain cadmium compounds.IARC monographs on the evaluation of the carcinogenic risk of chemicals to humans Beryllium, cadmium, mercury and exposures in the glass manufacturing industry IARC monographs. Lyon, France: World Health Organization 1993; Vol. 58: pp. 119-236.

[16]   Bernard AM, Lauwerys R. Effects of cadmium exposure in humans. Handb Exp Pharmacol 1986; 80: 135-77.
[http://dx.doi.org/10.1007/978-3-642-70856-5_5]

[17]   Suwazono Y, Kobayashi E, Okubo Y, Nogawa K, Kido T, Nakagawa H. Renal effects of cadmium exposure in cadmium nonpolluted areas in Japan. Environ Res 2000; 84(1): 44-55.
[http://dx.doi.org/10.1006/enrs.2000.4086] [PMID: 10991781]

[18]   Herrman JL, Younes M. Background to the ADI/TDI/PTWI. Regul Toxicol Pharmacol 1999; 30(2 Pt 2): S109-13.
[http://dx.doi.org/10.1006/rtph.1999.1335] [PMID: 10597623]

[19]   Renwick AG, Barlow SM, Hertz-Picciotto I, *et al.* Risk characterisation of chemicals in food and diet. Food Chem Toxicol 2003; 41(9): 1211-71.
[http://dx.doi.org/10.1016/S0278-6915(03)00064-4] [PMID: 12890421]

[20]   Tressou J, Crépet A, Bertail P, Feinberg MH, Leblanc JCh. Probabilistic exposure assessment to food chemicals based on extreme value theory. Application to heavy metals from fish and sea products. Food Chem Toxicol 2004; 42(8): 1349-58.
[http://dx.doi.org/10.1016/j.fct.2004.03.016] [PMID: 15207386]

[21]   Mild A, Reutterer T. An improved collaborative filtering approach for predicting cross-category purchases based on binary market basket data. J Retailing Consum Serv 2003; 10: 123-33.
[http://dx.doi.org/10.1016/S0969-6989(03)00003-1]

[22]   Scherer G, Renner T, Meger M. Analysis and evaluation of trans,trans-muconic acid as a biomarker for benzene exposure. J Chromatogr B Biomed Sci Appl 1998; 717(1-2): 179-99.
[http://dx.doi.org/10.1016/S0378-4347(98)00065-6] [PMID: 9832246]

[23]   Krishnamurti CR, Vishwanathan P. IEPHM, Toxic metals in the Indian Environment. New Delhi: Tata McGraw-Hill Publishing Company Ltd. 1991.

[24]   USEPA (US Environmental Protection Agency). 1996.

[25]   Asante-Duah DK. Risk Assessment in Environmental Management. England: John Wiley & Sons 1998.

[26]   Doubilet P, Begg CB, Weinstein MC, Braun P, McNeil BJ. Probabilistic sensitivity analysis using Monte Carlo simulation. A practical approach. Med Decis Making 1985; 5(2): 157-77.
[http://dx.doi.org/10.1177/0272989X8500500205] [PMID: 3831638]

[27]   Rabinowitz MB, Wetherill GW, Kopple JD. Kinetic analysis of lead metabolism in healthy humans. J Clin Invest 1976; 58(2): 260-70.
[http://dx.doi.org/10.1172/JCI108467] [PMID: 783195]

[28]   Azar A, Snee RD, Habibi K. An epidemiologic approach to community air lead exposure using personal air samplers. Environ Qual Saf Suppl 1975; 2: 254-90.
[PMID: 1058108]

[29]   McBride MB, Shayler HA, Spliethoff HM, *et al.* Concentrations of lead, cadmium and barium in urban garden-grown vegetables: the impact of soil variables. Environ Pollut 2014; 194: 254-61.
[http://dx.doi.org/10.1016/j.envpol.2014.07.036] [PMID: 25163429]

[30]   Agarwal SB, Singh A, Sharma RK, *et al.* Bioaccumulation of heavy metals in vegetables: A threat to human health. Terr Aquat Environ Toxicol 2007; 1(2): 13-23.

[31]   Zhou H, Yang WT, Zhou X, *et al.* Accumulation of heavy metals in vegetable species planted in contaminated soils and the health risk assessment. Int J Environ Res Public Health 2016; 13(3): 289.
[http://dx.doi.org/10.3390/ijerph13030289] [PMID: 26959043]

[32]   Al-Chaarani N, El-Nakat H, Obeid PJ, Aouad S. Measurement of levels of heavy metal contamination in vegetables grown and sold in selected areas in lebanon. Jordan Journal of Chemistry 2009; 4(3): 303-15.

[33]   Tripathi RM, Raghunath R, Krishnamoorthy TM. Dietary intake of heavy metals in Bombay city, India. Sci Total Environ 1997; 208(3): 149-59.
[http://dx.doi.org/10.1016/S0048-9697(97)00290-8] [PMID: 9496637]

[34]   Zhang ZW. Subida, R.D.; Agetano, M.G.; e al. Non-occupasional exposure of adult women in Manila, the Philippines, to lead and cadmium. Sci Total Environ 1998; 215: 157-65.
[http://dx.doi.org/10.1016/S0048-9697(98)00118-1] [PMID: 9599459]

[35]   Johnson RA. Miller, Freund. New Delhi: Probability and Statistics for Engineers, Prentice-Hall of India Private Limited 1994.

[36]   FAO/WHO (Food and Agriculture Organization/World Health Organization). 1999.

[37]   Sharma M. Air quality assessment, emissions inventory and source apportionment studies for Kanpur city. A report submitted to Central Pollution Control Board, New Delhi. 2010.

[38]   WHO. (World Health Organization) Recommended health-based limits in occupational exposure to heavy metals. Geneva: WHO 1980.

[39]   Forbes GB, Bruining GJ. Urinary creatinine excretion and lean body mass. Am J Clin Nutr 1976; 29(12): 1359-66.
[http://dx.doi.org/10.1093/ajcn/29.12.1359] [PMID: 998546]

[40]   ICRP (International Commission on Radiological Protection). Report of the task group on reference man. 1981.

[41]   Nordberg GF, Jin T, Kong Q, *et al.* Biological monitoring of cadmium exposure and renal effects in a population group residing in a polluted area in China. Sci Total Environ 1997; 199(1-2): 111-4.
[http://dx.doi.org/10.1016/S0048-9697(97)05486-7] [PMID: 9200853]

[42]   Staessen J, Amery A, Bernard A, *et al.* Effects of exposure to cadmium on calcium metabolism: a population study. Br J Ind Med 1991; 48(10): 710-4.
[http://dx.doi.org/10.1136/oem.48.10.710] [PMID: 1931731]

[43]   Sartor FA, Rondia DJ, Claeys FD, *et al.* Impact of environmental cadmium pollution on cadmium exposure and body burden. Arch Environ Health 1992; 47(5): 347-53.
[http://dx.doi.org/10.1080/00039896.1992.9938373] [PMID: 1444596]

[44]   van Sittert NJ, Ribbens PH, Huisman B, Lugtenburg D. A nine year follow up study of renal effects in workers exposed to cadmium in a zinc ore refinery. Br J Ind Med 1993; 50(7): 603-12.
[http://dx.doi.org/10.1136/oem.50.7.603] [PMID: 8343421]

[45]   Taylor SA, Chivers ID, Price RG, *et al.* The assessment of biomarkers to detect nephrotoxicity using an integrated database. Environ Res 1997; 75(1): 23-33.
[http://dx.doi.org/10.1006/enrs.1997.3775] [PMID: 9356191]

[46]   Staessen JA, Lauwerys RR, Ide G, Roels HA, Vyncke G, Amery A. Renal function and historical environmental cadmium pollution from zinc smelters. Lancet 1994; 343(8912): 1523-7.
[http://dx.doi.org/10.1016/S0140-6736(94)92936-X] [PMID: 7911869]

[47]   Roels H, Bernard AM, Cárdenas A, *et al.* Markers of early renal changes induced by industrial pollutants. III. Application to workers exposed to cadmium. Br J Ind Med 1993; 50(1): 37-48.
[http://dx.doi.org/10.1136/oem.50.1.37] [PMID: 8431389]

[48]   Jung K, Pergande M, Graubaum HJ, Fels LM, Endl U, Stolte H. Urinary proteins and enzymes as early

indicators of renal dysfunction in chronic exposure to cadmium. Clin Chem 1993; 39(5): 757-65.
[http://dx.doi.org/10.1093/clinchem/39.5.757] [PMID: 7683580]

[49]   Zhang ZW, Shimbo S, Watanabe T, *et al.* Non-occupational lead and cadmium exposure of adult women in Bangkok, Thailand. Sci Total Environ 1999; 226(1): 65-74.
[http://dx.doi.org/10.1016/S0048-9697(98)00370-2] [PMID: 10077875]

[50]   Higashikawa K, Zhang ZW, Shimbo S, *et al.* Correlation between concentration in urine and in blood of cadmium and lead among women in Asia. Sci Total Environ 2000; 246(2-3): 97-107.
[http://dx.doi.org/10.1016/S0048-9697(99)00415-5] [PMID: 10696716]

CHAPTER 6

Safety Evaluation of Coloured Plastic Tiffins/Bottles and Medical Strategies to Mitigate Additive Toxicity

Sonika Bhatia[1,*]

[1] *Isabella Thoburn College, Lucknow, India*

Abstract: Over the last 50 years, plastics have become an integral part of our day-t--day life. They seem to be the material of choice, being inexpensive, lightweight, resistant to chemicals, flexible, and mouldable. Over 350 million metric tons of plastics are produced the worldwide and about 50% are discarded within the first year of usage. With plastics being non-biodegradable, their safe disposal is a major challenge. Although one can argue that it is recyclable, its indiscriminate disposal clogs up our rivers, oceans, and land. Some of the chemical components of plastics such as plasticizers, stabilizers, monomers, and colourants are known to leach out from the finished plastics into the stored commodity. They can also be released during various recycling processes. In the present study, undergraduate students were involved in analysing a few representative samples for the estimation of the overall leaching of plastic constituents and heavy metals into food and drinking water. Toxic effects of heavy metals, phthalates and BPA, probable mechanism of toxicity, and some medical strategies have been discussed. Findings from a survey carried out by the students to gauge awareness about leaching in plastics, segregation, and disposal of plastic wastes practised by the community are presented. This experiential learning is aimed at inculcating behavioural change about the judicious usage and proper waste disposal of plastics.

Keywords: BPA, Heavy metals, Judicious usage, Leaching, Medical strategies, Phthalates, Safety Evaluation, Toxicity.

INTRODUCTION

The past hundred and fifty years have witnessed an unbelievable pace of scientific and technological developments. Modern science has attempted and to a great extent succeeded in realizing the dreams of the scientists of the previous centuries. Ranging from the packaging of drinking water, foodstuff, and pharmaceuticals to biomedical implants, kitchenware, tabletops and alternate building materials,

* **Corresponding author Sonika Bhatia:** Department of Chemistry, Isabella Thoburn College, Lucknow, India; E-mail: sonikaitc2@gmail.com

Tahmeena Khan, Abdul Rahman Khan, Saman Raza, Iqbal Azad and Alfred J. Lawrence (Eds.)

plastics and polymeric products have become an integral part of our day-to-day life. They are materials of choice due to their inertness in the finished state, lightweight, high strength, amenability for quick and mass production, and ease of fabrication into complex shapes in a variety of colours. Over 350 million metric tons of plastics are produced annually worldwide and about 50% of them are discarded within the first year of their usage [1]. The chemical structure of plastics makes them resistant to various natural degradation processes. These two factors together have led to a huge accumulation of plastic waste in the world. Almost all water bodies form a web and eventually flow into the oceans. Thus, these rich natural resources act as conveyor belts for our trash. An overview of plastic toxicology is given in Fig. (**1**).

PLASTIC WASTE POSES A HUGE THREAT TO MARINE ECOLOGY			
There are 8 million tonnes of plastic waste entering the ocean every year	The total plastic in the ocean amounts to 150 million tonnes	Plastic packaging accounts for 62% of all items recovered in coastal clean-up efforts	In 2014, there was 1 Kg of plastic in the ocean for every 5 Kg of fish, and by 2050 there will be more plastic than fish

Fig. (1). Plastic toxicology- An overview [2].

As most of the floating plastic waste is non-biodegradable, it does not decompose and keeps floating for years leading to overtime lowering of oxygen level in the water. Moreover, when marine creatures and birds accidentally consume plastic waste, they choke on it, thus leading to a steady decline in their population. Another worrying aspect of plastic usage is the migration of additives from plastic containers into the stored commodity. Most of the additives like plasticizers (*viz.* Phthalate Esters), colourants and stabilisers (*viz.* organo-tin, lead, and barium compounds) added to plastics to improve their physicochemical properties are not chemically bound to the polymers. These being small, are liable to dissolve in the aqueous environment and leach out over time from the containers into the foodstuff and water stored in them. They can also be released from the plastics during various recycling processes and from the products manufactured from recyclates. Various studies [3 - 5] have shown that heavy metals leach out from the plastics and get accumulated in the human body and additives like Di(2-Ethylhexyl)phthalate(DEHP) and Bisphenol A (BPA), which are toxic, have been found in traces in human blood, urine and sweat [6, 7].

Due to a large demand for inexpensive brightly coloured plastic tiffin boxes, bottles, and toys for children, several small-scale industries are involved in manufacturing these plastic products from both fresh and recycled plastics. Often to make the articles attractive, additives, more than the prescribed limit, are added to attain the desired quality. In India, 60-65% of the recycling industry falls in the unorganised sector, thus fewer or no quality checks are performed before the finished products are sold in the market.

Children carry tiffin boxes and bottles to school daily; often hot food is packed in these tiffin boxes and the water bottles are left in the sun. The overtime leaching of additives could be detrimental to their health. DEHP and BPA are well-known endocrine disruptors. Since hormones control and coordinate activities throughout the body, even a small disruption can interfere with the child's growth and development. A brief review of the toxic effects of common migrating additives and some medical interventions to mitigate additive toxicity is presented here.

HEAVY METAL TOXICITY

Some heavy metals like zinc, copper, chromium, iron, and manganese are naturally found in our body and are necessary for regular body functioning but when their levels cross the critical limit [8] in the body tissues, toxicity is observed (Fig. **2**). These elements gain access into our bodies *via* food, air, or water. General symptoms of mild metal toxicity can be diarrhoea, nausea, abdominal pain, shortness of breath, weakness, and increased chances of miscarriage. Copper is essential for haemoglobin formation and carbohydrate metabolism. In excess, however, it causes cellular damage [9]. Exposure to very high levels of lead in children has been found to decrease attention span, increase irritability, and lower intelligence [10]. Further, lead is reported to be primarily stored in the bones and teeth which accumulate over time to affect the bones, kidney, and liver [11]. Long-term exposure may result in degenerative changes leading to multiple sclerosis, Parkinson's disease, muscular dystrophy, and Alzheimer's disease.

TOXICITY MECHANISM

The majority of heavy metals facilitate the formation of ROS (Reactive oxygen species/free radicals) and reduce the level of antioxidants like glutathione in the body leading to oxidative stress in the body (Fig. **2**). This increased level of oxygen free radicals in the body may lead to lipid peroxidation [12] and structural damage of cells, proteins, and nucleic acids [13]. Another pathway for toxicity is when the metal cations present in the biological systems are displaced by heavy metal pollutant cations; this may lead to an interruption in many biological processes like cell adhesion, intra and intercellular signalling, enzyme regulation, apoptosis, protein folding, ionic transportation, and release of neurotransmitters [13, 14].

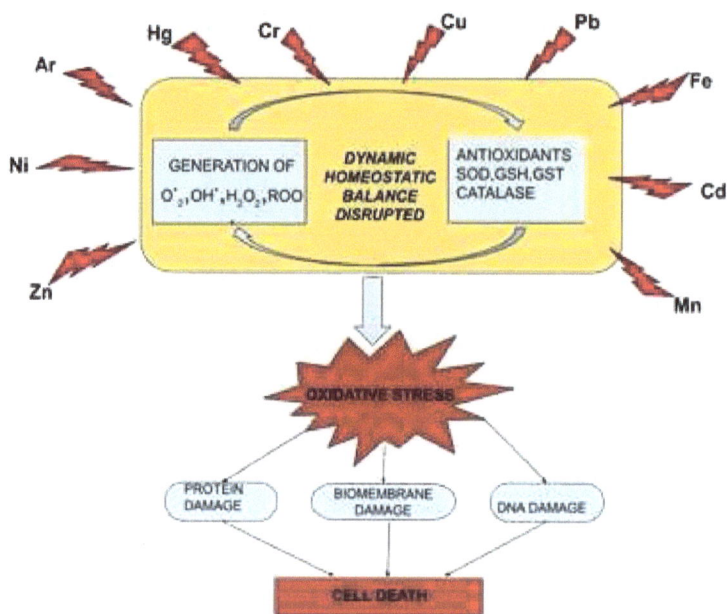

Fig. (2). Mechanism of heavy metal toxicity.

MEDICAL INTERVENTIONS AND STRATEGIES

Chelation therapy is the preferred medical intervention for reducing the toxic effect of metals. Chelation is a process in which ions/molecules of a ligand bind to the central metal atom/ions *via* coordinate bonds, in a cyclic or ring-like structure. A good chelating agent is one whose chemical affinity for the metal ions is higher than the affinity of the metal ions for the sensitive biological molecules. It should be stable at the pH of body fluids, be nontoxic, and should have appropriate pharmacokinetic properties to be easily administered (orally, intravenously, or intramuscularly) and easily excreted from the body.

COMMON CHELATING AGENTS IN USE

Some common chelating agents (Fig. 3) have been described below.

DIMERCAPROL

It is also referred to as British Antilewisite (BAL), and it was developed to treat exposure to the arsenic-based poisonous gas Lewisite, during the Second World War. Since then, dimercaprol has been used effectively to treat arsenic, mercury, lead, and gold poisoning. The two thiol groups in dimercaprol form a stable 5-membered metal-ligand ring complex which is excreted through the kidneys. Sweating, fever, hypertension, headache, nausea, vomiting, and palpitation are some of the side effects associated with dimercaprol treatment. Moreover, it can only be administered through the parenteral route being water-insoluble [15].

DMSA & DMPS-ORAL CHELATION ANALOGUES OF DIMERCAPROL

Meso-2, 3–dimercaptosuccinic acid (DMSA), also known as Succimer is also a dithiol like BAL but is more water-soluble and thus has fewer side effects. DMSA contains two carboxylic groups and two thiol groups. Like Dimercaprol, the two thiol groups are involved in the formation of the chelation complex. DMSA has been effectively used to treat lead, mercury, and arsenic toxicity, especially in children [16, 17]. 3-Dimercapto-Propanesulphonate (DMPS) is another water-soluble analogue of dimercaprol. DMPS contains one sulfonic group and two thiol groups. Both these drugs have been consistently used for lead, mercury, and arsenic toxicity with few temporary side effects like skin reactions, mild neutropenia, and moderately elevated liver enzymes [18].

SODIUM-CALCIUM EDTA

The calcium atom of this compound is easily replaced by other metal ions to form a water-soluble complex that is eliminated *via* kidneys. $CaNa_2EDTA$ can remove the lead accumulated in the soft tissues but is unable to cross the blood-brain barrier effectively [19]. Endogenous metals like zinc, copper, and iron also get excreted out to some extent. Thus, there is a need for supplementation of these essential metals [20].

DEFEROXAMINE (DFO)

It contains three-hydroxamic acid groups which can form strong bonds with trivalent metal cations. DFO is specifically used for iron and aluminium toxicity. It forms ferrioxamine and aluminoxamine complexes which are easily eliminated through kidneys [16, 21].

PENICILLAMINE AND TRIENTINE (TRITHYLENETETRAMINE)– ORAL COPPER CHELATORS

These compounds are mainly used to tackle copper poisoning. However, the clinical use of D- penicillamine is limited due to its side effects, *i.e.*, nephrotic syndrome and various autoimmune reactions which develop in a considerable fraction of patients during its continuous use. Trientine chelates divalent copper in the body by using nitrogen atoms as electron donor groups. This chelator has fewer side effects in comparison to penicillamine and can be given by oral route [22, 23].

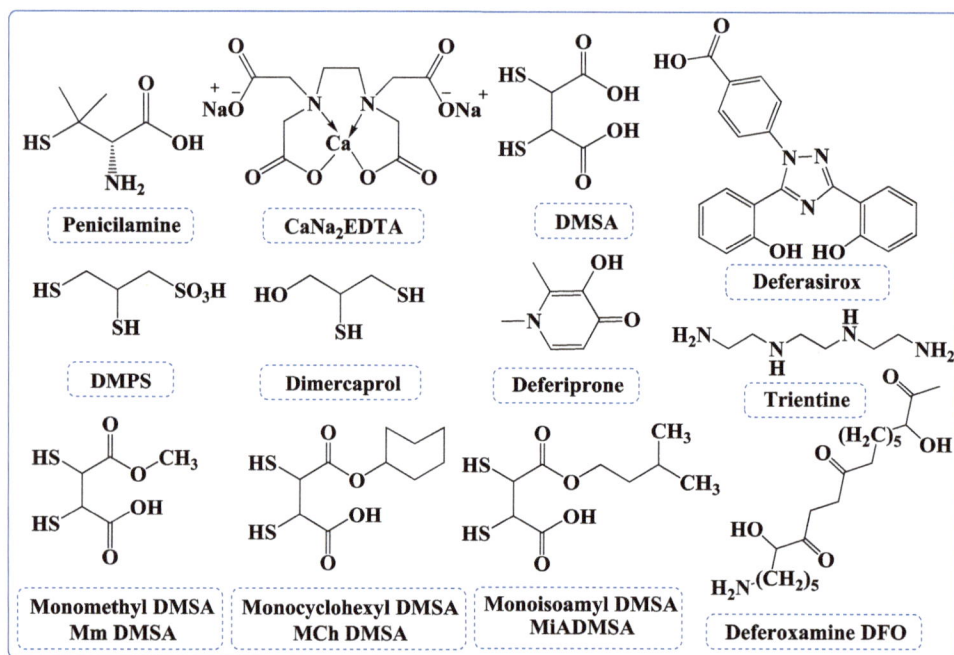

Fig. (3). Common chelating agents.

DMSA ANALOGUES

Many synthetic analogues of DMSA have been developed like monoisoamyl DMSA (MiADMSA), monocyclohexyl DMSA (MchDMSA) and monomethyl DMSA (MmDMSA). These analogues, being more lipophilic than DMSA, are more effectively able to penetrate the cell membranes and remove heavy metals. They also showed better excretory efficacy than DMSA. These analogues have been used alone or in combination, to reduce toxicity due to arsenic, cadmium, and mercury [15, 24, 25].

DEFERIPRONE AND DEFERASIROX-ORAL IRON CHELATORS

These are both orally active tridentate chelators that have been successfully used to treat iron and aluminium toxicity. For many years Deferoxamine (Desferal) was the drug of choice for iron toxicity. As it is poorly absorbed by the gastrointestinal tract, it must be administered by intravenous or subcutaneous route. Water-soluble chelators Deferiprone and Deferasirox [15] can easily be given orally.

COMBINATION THERAPY

A combination of $CaNa_2EDTA$ and MiADMSA has been reported to show better biochemical and clinical results than monotherapy [26]. Studies have shown that α-lipoic acid in combination with thiol chelators reduces oxidative stress and brain lead concentration in rats [27]. A combination of antioxidants (such as Vitamin E and C, thiol group, zinc, and selenium) and chelating agents works well by reducing oxidative stress. This treatment helps to reduce side effects leading to better recovery [15].

PHTHALATE TOXICITY

Phthalates are frequently added to plastics as plasticizers to create plastic products that are soft and malleable. Some of the phthalates which are commonly used are Di-2-ethyl-hexyl phthalate (DEHP), Diisononyl phthalate (DINP), and Diisodecyl phthalate (DIDP). DEHP is mostly added to polyvinyl chloride. Once the phthalates enter the body, they undergo phase I hydrolysis and phase II conjugation metabolic reactions and are readily excreted *via* the kidneys. Nevertheless, phthalates have been frequently detected in the blood, urine, and sweat samples of individuals, suggesting rampant and continuous human exposure to phthalates [6]. DEHP, being lipophilic, has been found in various high-fat content food products like meat, oils, and dairy products in $\geq 300\mu g/kg$ concentrations [28, 29]. According to the Environmental Protection Agency (EPA), a dose higher than 20 µg/kg/day can be toxic and may lead to endocrine and testicular toxicity [28]. DEHP is a known endocrine disrupter [30] and is known to induce oxidative stress leading to insulin resistance in the elderly [31]. Studies have pointed out that DEHP disturbs the thyroid hormone homeostasis by reducing T3, T4, and TRH (Thyrotropin-releasing hormone) [32]. Studies have also shown that DEHP is linked to obesity, especially in children [33, 34]. DEHP has been associated with ovarian, renal, hepatic, cardio- and neurotoxicity [35] as well as endometriosis in women [36].

BISPHENOL A TOXICITY

Bisphenol A (BPA) is found in a lot of plastic consumer products ranging from plastic containers, bottles, pipes, bags, toys, medical equipment, *etc*. It is used as a monomer in polycarbonate synthesis, as a plasticizer in epoxy resin synthesis, and as a stabilizer in polyvinyl chloride production. A study conducted by the Centre of Disease Control USA [37] has shown that 92.6% of urine samples tested contained BPA in varying concentrations, with children showing higher concentration (4.5µg/l) as compared to adults (2.5 µg/l). Bisphenol A is a known endocrine disrupter [38]. Its interaction with estrogen receptor, androgen receptor, and thyroid hormone receptor leads to several endocrine disorders, including female and male infertility, precocious puberty, hormone-dependent tumours such as breast and prostate cancer, and several metabolic disorders including polycystic ovary syndrome (PCOS) [39, 40]. With the widespread concern over BPA toxicity, governments across the world have decreased the permissible limit of BPA and have put a blanket ban on the use of BPA in infant feeding bottles [41, 42]. These restrictions have paved the way for the development of many BPA analogues like Bisphenol-S(BPS){bis(4-hydroxyphenyl)sulfone}, Bisphenol F(BPF){bis(4-hydroxyphenyl)methane}, Bisphenol-B(BPB){2,2-bis-4-hydroxyphenyl)butane} and Bisphenol-AF (BPAF) {4,4-(hexafluoroisopropylidene)diphenol}. Due to their structural similarities with BPA, they have also been found to disrupt the endocrine system [43]. BPA leaching seems to increase with the contact time, temperature, and pH [44], and canned or possessed foods are found to contain higher levels of BPA [45]. Though oral ingestion is the main route of BPA exposure in humans, studies have shown that BPA exposure can also occur through dermal contact [46]. Thermal papers used in adding machines, ticketing counters, and ATMs are coated with BPA which function as a heat-activated developer for the dyes used. This BPA coating is not chemically bound thus easily gets transferred to the skin on normal handling [47]. When BPA enters the body through the oral route, most of it easily gets metabolised to water-soluble conjugates: BPA Glucuronide and BPA Sulphate. These are easily excreted from the body. The endocrine disruption activity is due to the unmetabolised leftover BPA in the body. But when BPA is absorbed from the dermal surface, it escapes the first pass mechanism, and the unconjugated BPA circulates in the bloodstream [48] resulting in higher toxicity. Though it may be argued that the exposure to these thermal slips is a bare minimum, people on cash desks and ticketing facilities are at a higher risk [49].

THERAPEUTIC STRATEGIES TO REDUCE TOXICITY DUE TO PHTHALATES AND BPA

The best strategy seems to be

- Reducing exposure to phthalates and BPA.
- Increasing the rate of elimination from the body.
- Building up the antioxidant level in the body to mitigate oxidative stress triggered by phthalate and BPA overload.

REDUCING EXPOSURE

Exposure to additives can be minimised by using glass, ceramic, or metal containers to store food or warm foods. As phthalates and BPA are lyophilic, fat-rich foods like meat, oil, and dairy products should not be stored in plastic containers. Studies have confirmed that phthalate and BPA levels in food increase with an increase in storage time, consequently concentration of DEHP is more in packaged food items closer to their expiry date [4]. Thus, food bought in plastic containers should be transferred to glass containers immediately and food items closer to the expiry date should not be consumed. Moreover, as the rate of migration increases with temperature, hot oily food should not be packed in plastic tiffin boxes/containers and hot liquids should not be stored in plastic bottles. Increased migration of DEHP has been shown in plastic bottles exposed to the sun or left in closed cars [50].

INCREASING RATE OF ELIMINATION

As phthalates get metabolised to water-soluble metabolites and are easily excreted out of the body *via* urine and sweat [6, 7], drinking 2.5-3 litres of water daily and inducing precipitation by strenuous exercise/sauna baths to flush out the toxins from the body, might be a good strategy. The use of a high fibre diet can also facilitate detoxification [51].

ANTIOXIDANT SUPPLEMENTS

Supplementing the diet with antioxidants like Vitamin E and C and natural fruits and vegetables that are rich in antioxidants, can mitigate oxidative stress catalysed by an overload of phthalates and BPA in the body [52].

Studies have shown that natural antioxidants like Curcumin (the active component of turmeric) and Resveratrol (3, 4, 5 trihydroxy-trans-stilbene) found in mulberries, grapes, blueberries, and cranberries are very effective in tackling DHEP toxicity [53]. Lycopene has been reported to reduce the cytotoxic effects of BPA on the hepatic tissue in rats [54]. Protective effects of naturally occurring coenzyme Q10 against BPA toxicity have also been studied [55, 56].

A CASE STUDY

In a bid to provide experiential learning to undergraduate students, they were involved in analysing a few representative samples of coloured tiffin boxes and water bottles procured from the local market, for the estimation of overall leaching of plastic constituents and heavy metals such as Cd, Cr, Cu, Fe, Mn, Ni, Pb, and Zn into foodstuff and drinking water according to Bureau of Indian Standards (BIS) guidelines [8]. The permissible limit for various heavy metals is 1 ppm and that for cadmium is 0.1 ppm and migration residues should not be more than 5mg/100 ml of the simulant. A survey was carried out by the students to gauge the awareness about the leaching from plastics into stored food and the disposal of plastic wastes practised by the community. This experiential learning was aimed at inculcating behavioural change about judicious usage and proper disposal of plastic waste. The study was done with a three-pronged approach:

- Assessing Susceptibility
- Correlation *via* a survey
- Awareness Building

RESULTS

Assessing Susceptibility-Global migration and estimation of heavy metals (Cd, Cr, Cu, Fe, Mn, Ni, Pb, and Zn) tests were conducted at 25°C and 60°C, as per IS 985:1998 guidelines, on 20 samples in 5% Na_2CO_3, 0.9% NaCl, 8% C_2H_5OH, 3% CH_3COOH and double distilled water used as simulating solvents (Figs. **4 - 11**). Permissible limit: Global migration residues should not be more than 5mg/100 ml of the simulant.

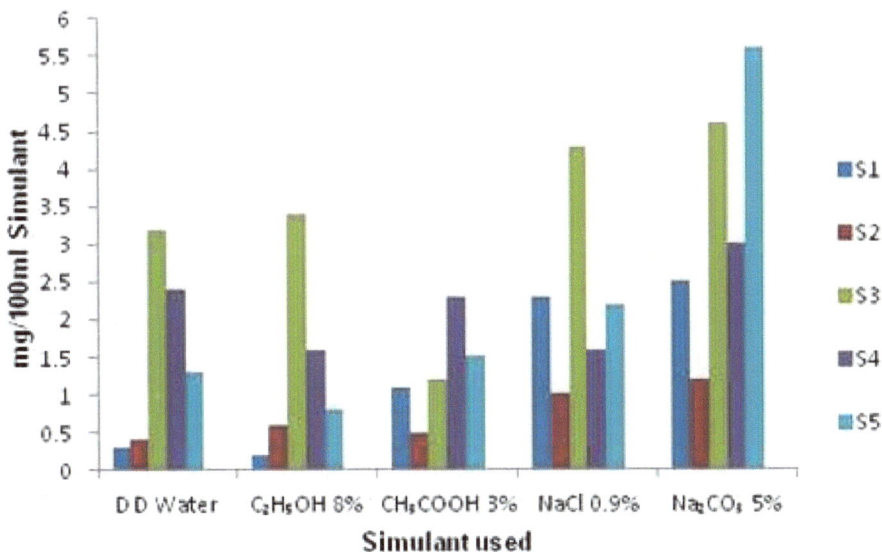

Fig. (4). Global migration residues (mg/100ml) of the simulants of coloured tiffin boxes/bottles at 25°C +/-2°C for 24 hours.

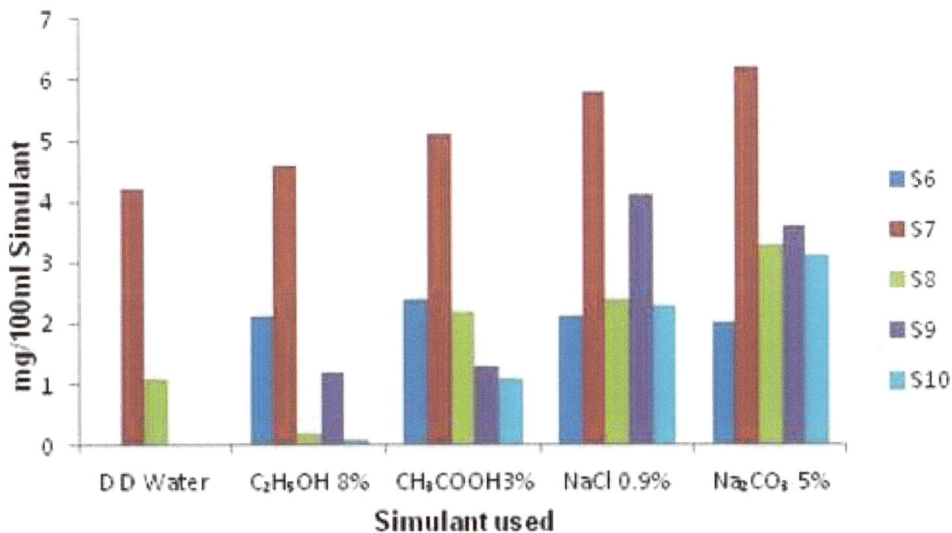

Fig. (5). Global migration residues (mg/100 ml) of the stimulants of coloured tiffin boxes/bottles at 25 ^{0}C. +/- 2 ^{0}C for 24 hours.

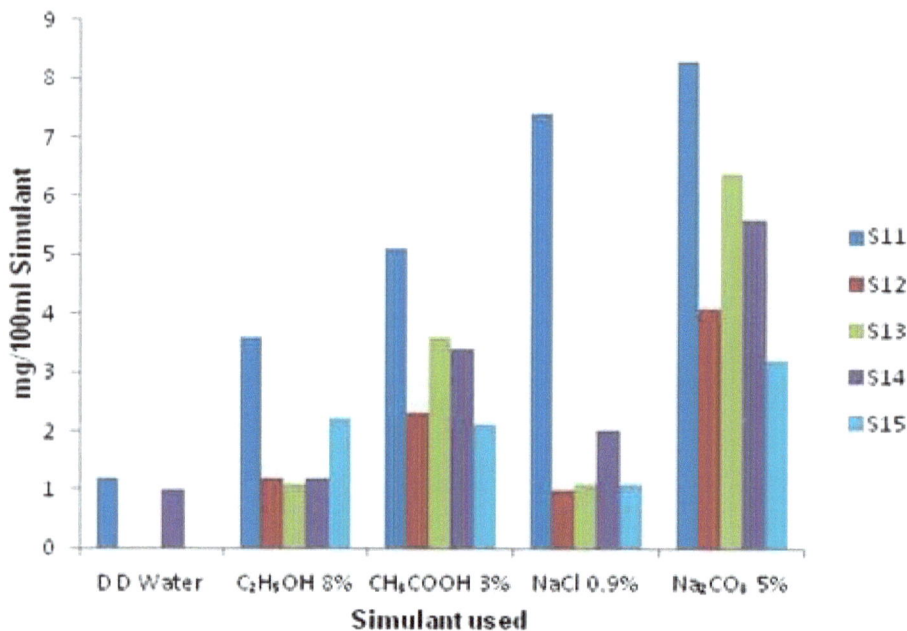

Fig. (6). Global migration residues (mg/100ml) of the simulants of coloured tiffin boxes/bottles at 25 ^{0}C.

+/-2 ^{0}C for 24 hours.

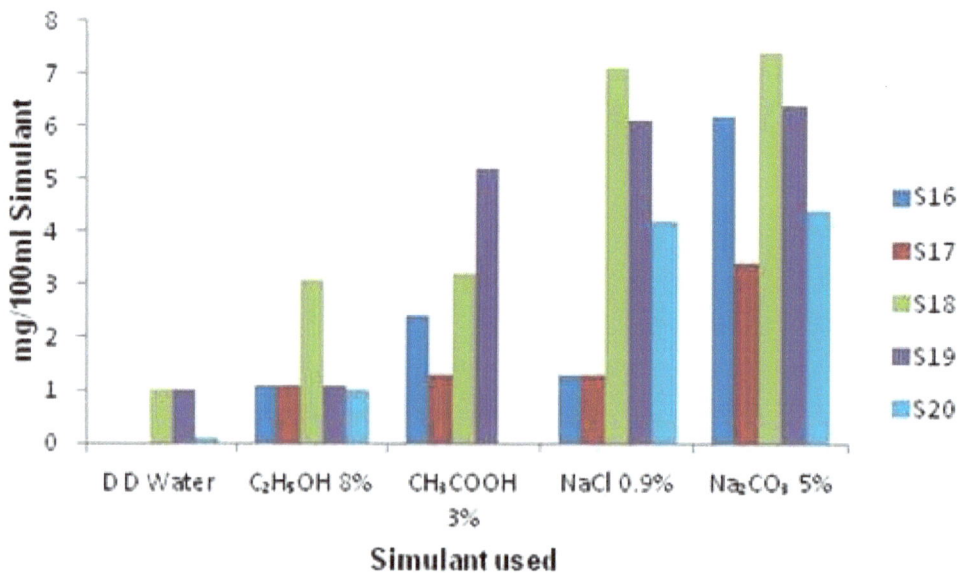

Fig. (7). Global migration residues (mg/100ml) of the simulants of coloured tiffin boxes/bottles at 25 ^{0}C. +/-2 ^{0}C for 24 hours.

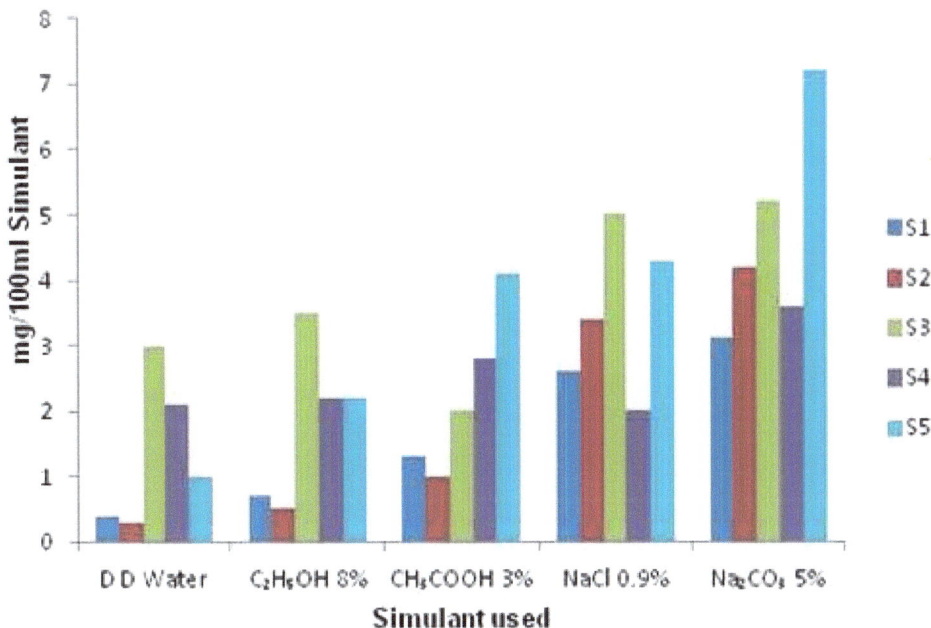

Fig. (8). Global migration residues (mg/100ml) of the simulants of coloured tiffin boxes/bottles at 60 ^{0}C.

+/-2 ^{0}C for 24 hours.

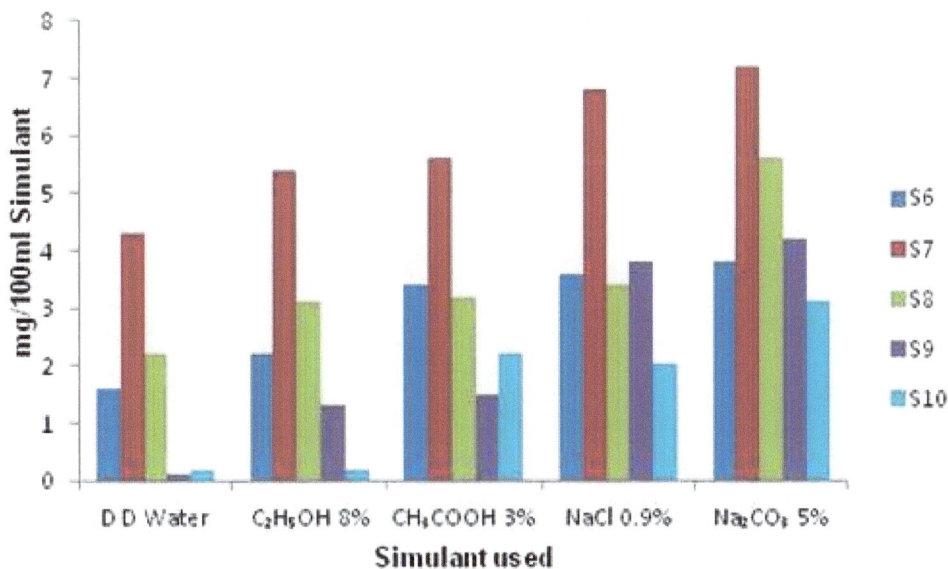

Fig. (9). Global migration residues (mg/100ml) of the simulants of coloured tiffin boxes/bottles at 60 ^{0}C.

+/-2 ^{0}C for 24 hours.

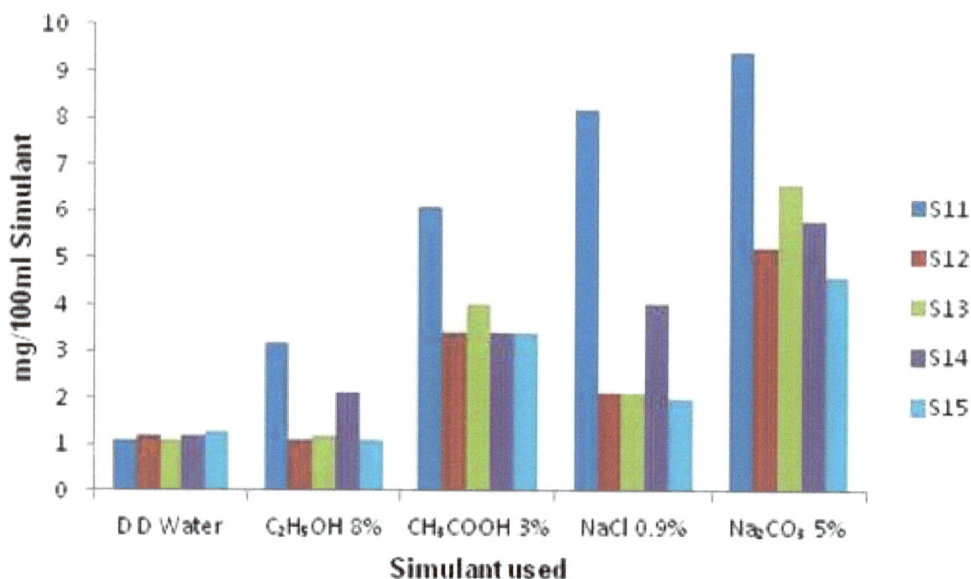

Fig. (10). Global migration residues (mg/100ml) of the simulants of coloured tiffin boxes/bottles at 60 $^{\circ}$C.

+/-2 $^{\circ}$C for 24 hours.

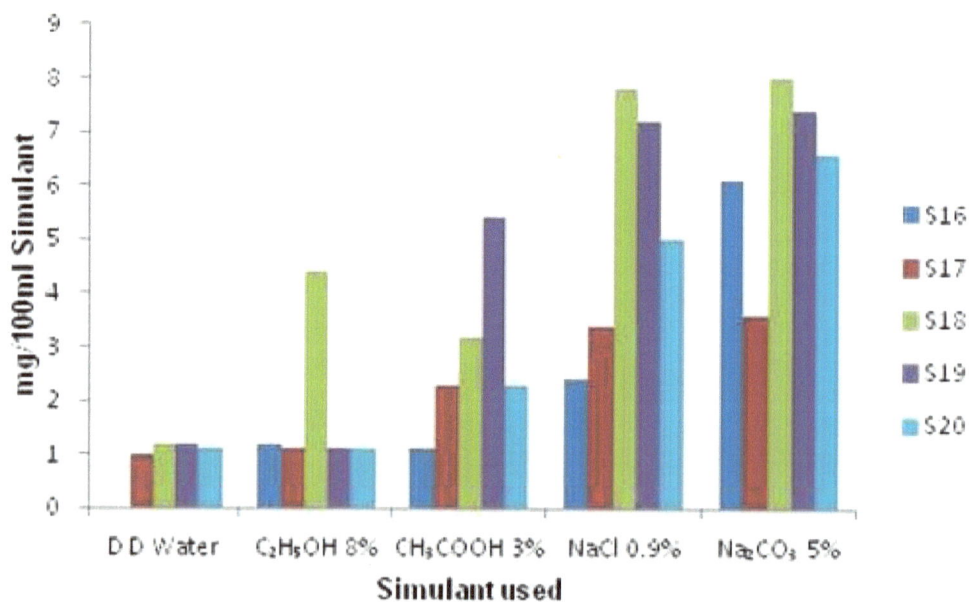

Fig. (11). Global migration residues (mg/100ml) of the simulants of coloured tiffin boxes/bottles at 60 $^{\circ}$C.

+/-2 $^{\circ}$C for 24 hours.

HEAVY METAL MIGRATION (IN PPM)

Permissible limit: Cr, Cu, Fe, Mn, Ni, Pb, Zn in plastic extract should not exceed 1 ppm and that of Cd should not exceed 0.1 ppm. The findings indicated that the leaching of heavy metals in samples was temperature, as well as solvent dependent. Out of the 20 samples analysed, leaching of metals was found to be within the prescribed permissible limits for Zn, Cu, Fe, Mn, and Ni (Figs. **12 - 18**). Only in sample 11(Yellow), the leachates of 5% Na_2CO_3 at 60 ^{0}C +/-2 ^{0}C for 2 hours showed the presence of lead at 1.2 ppm and 8% ethanol leachates at 60 $^\circ$C +/-2 ^{0}C for 2 hours exposure showed the presence of cadmium at 0.2 ppm. Sample14(Red) showed the presence of lead at 1.2 ppm in the 0.9% NaCl leachate at 25 ^{0}C +/-2 for 24 hours exposure, while sample 16 (Green) exhibited the presence of chromium at 1.01 ppm in 0. 9% NaCl leachate at 25 ^{0}C +/-2, for 24 hours exposure.

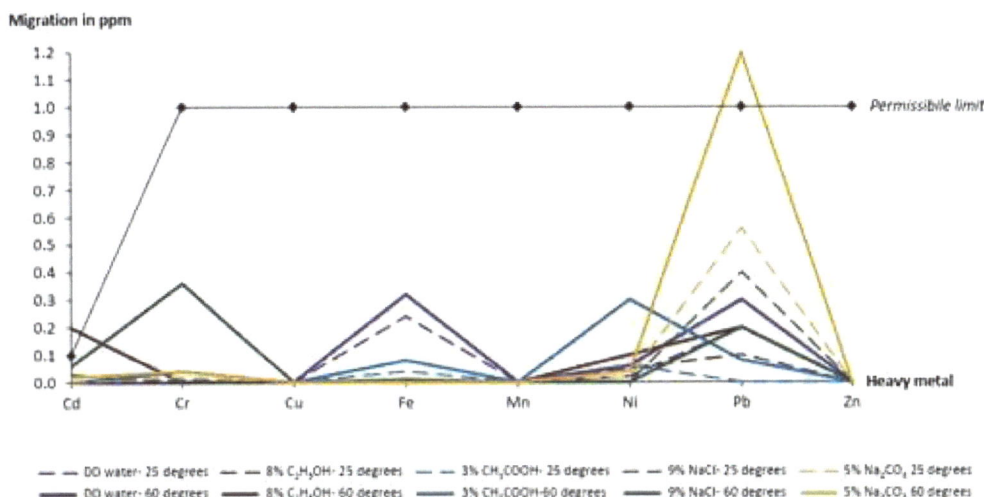

Fig. (12). Migration of heavy metals in sample 11(in ppm).

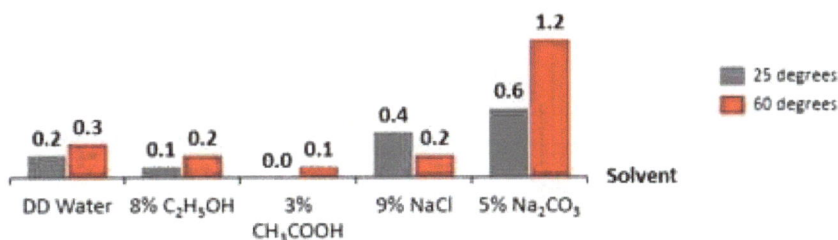

Fig. (13). Migration of Lead in sample 11(in ppm).

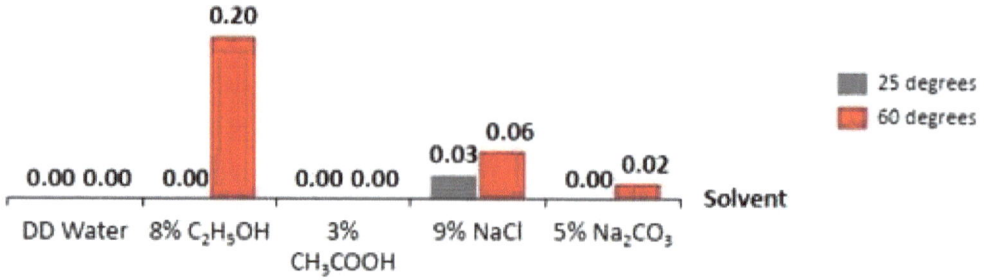

Fig. (14). Migration of Cadmium in sample 11(in ppm).

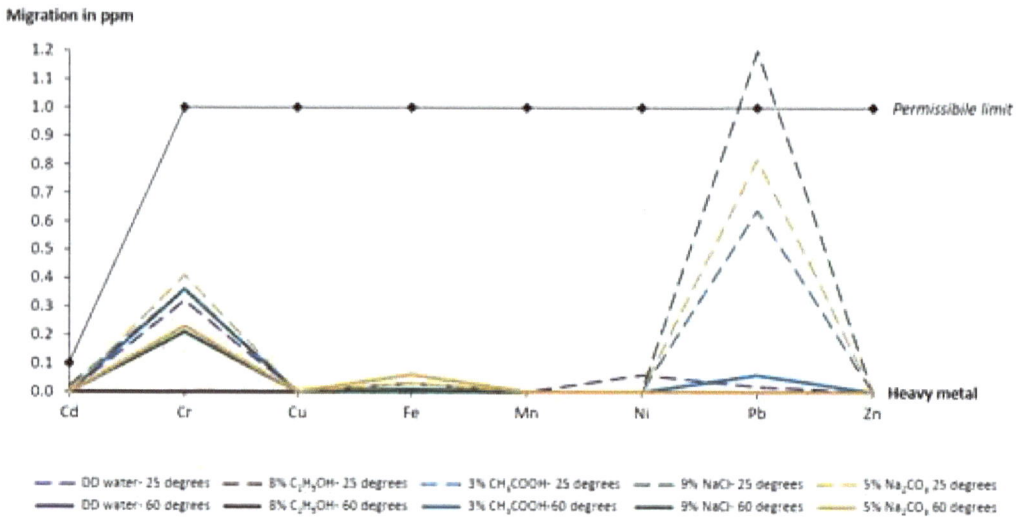

Fig. (15). Migration of heavy metals in sample 14(in ppm).

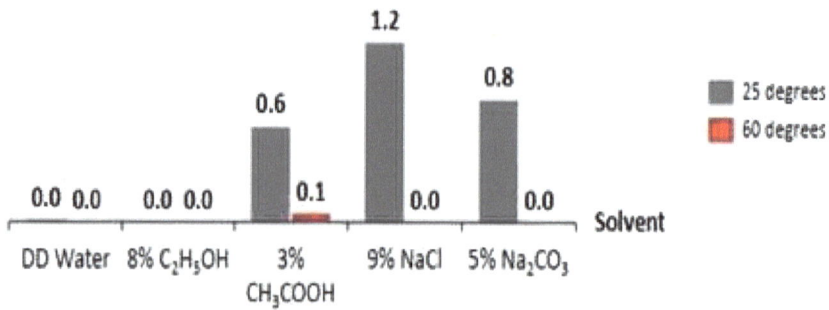

Fig. (16). Migration of Lead in sample 14(in ppm).

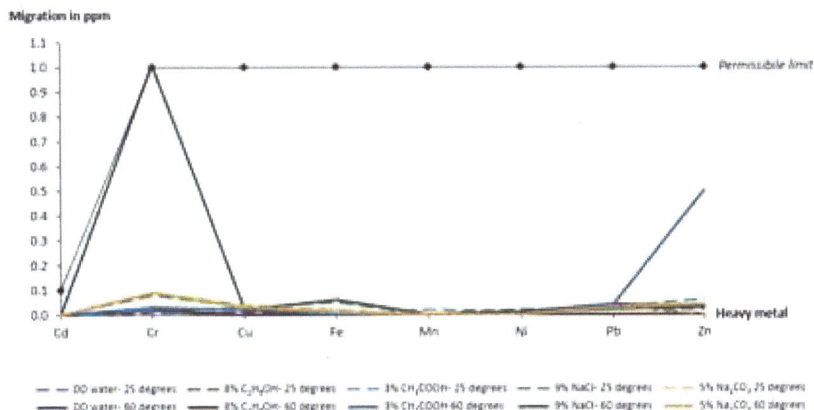

Fig. (17). Migration of heavy metals in sample 16(in ppm).

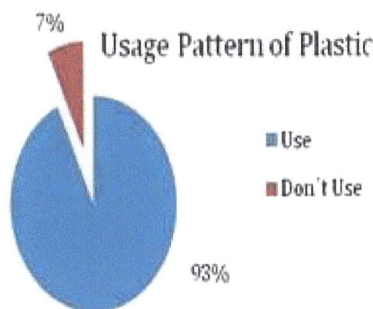

Fig. (18). Migration of Chromium in sample 16(in ppm).

SURVEY

A pilot survey was conducted covering 463 households in the city of Lucknow, to ascertain the household pattern of plastic usage and knowledge about leaching of colourants and additives into food, among the people of Lucknow.

OBJECTIVES

- To find out the % of children carrying coloured plastic bottles and tiffin boxes to school.
- To find out the kind of bags usually used by people to buy vegetables, pickles, curd, and grocery.
- To ascertain the usual pattern of disposal of plastic and biomedical waste.
- To ascertain the percentage of people practising segregation of garbage.
- To ascertain the level of awareness about the harmful effects of plastics on animals & marine life.

TOOLS USED

A short questionnaire was used to ensure the involvement of both the interviewer and the respondent throughout the survey. All the interviews were conducted on a one-to-one basis.

INTERPRETATION AND ANALYSIS OF SURVEY

The responses to the questionnaire were tabulated for analysis and are shown in Figs. (**19 - 29**) .

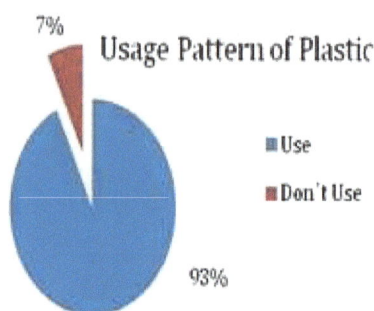

Fig. (19). Response to- 'Do you use plastics or plastic products?' shows high usage of plastics in urban areas.

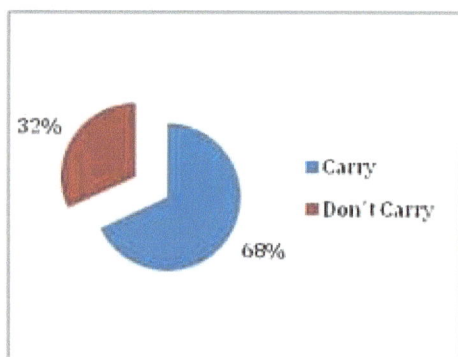

(a)

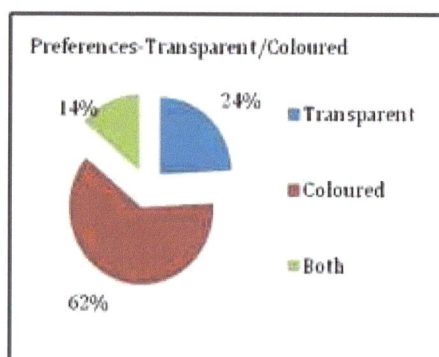

(b)

Fig. (20). (a), Response To-Do your children carry plastic tiffin boxes/bottles to school? **(b)**, If yes, coloured /transparent?.

Fig. (**20**) shows that 68% of children carried plastic bottles and tiffin boxes while 32% of children carried steel tiffin boxes or no tiffin at all. Thus, it was concluded that 0.68 x 0.76=51.7% of children were at health risk as warm foods/drinks are being carried in coloured containers leading to higher chances of leaching.

(a)

(b)

(c)

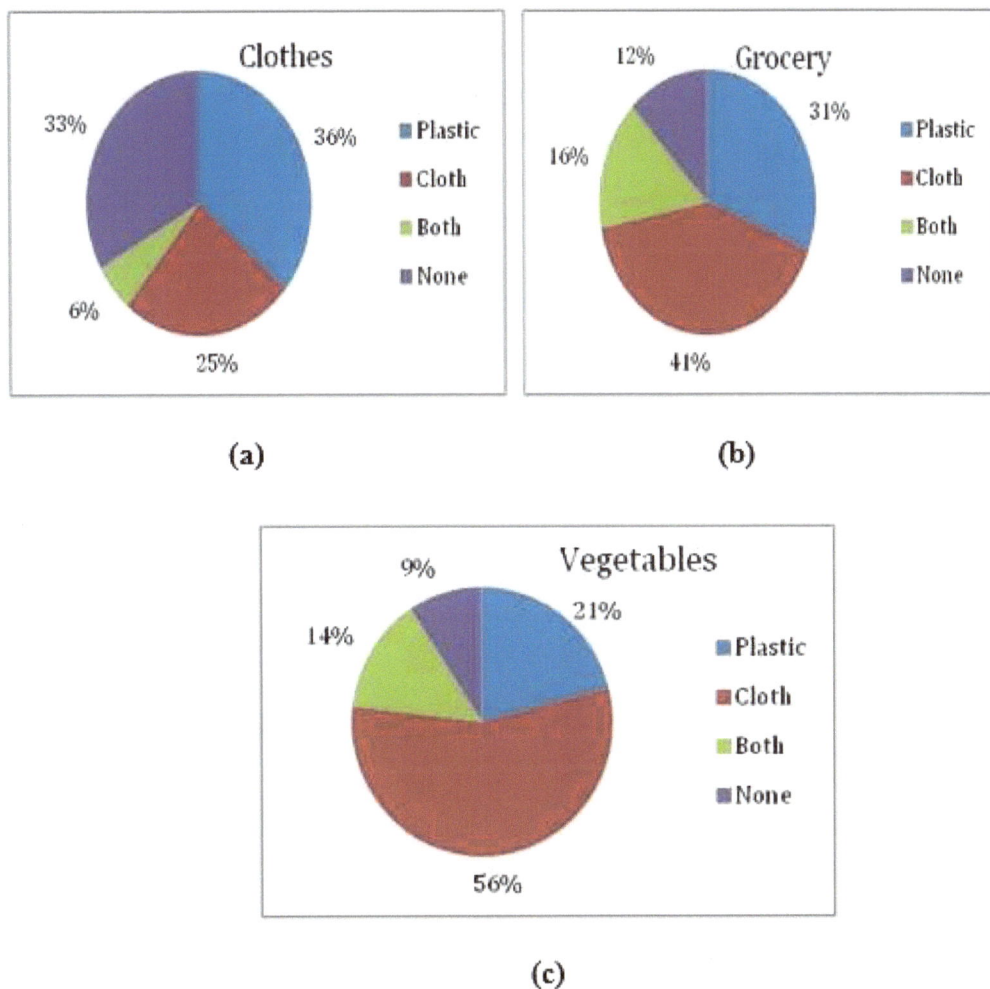

Fig. (21). Response to- when you go shopping, what kind of bags do you carry? **(a)**, clothes, **(b)**, grocery, **(c)**, vegetable.

Fig. (**21**) shows high usage of plastic bags while shopping for vegetables, grocery, and clothes.

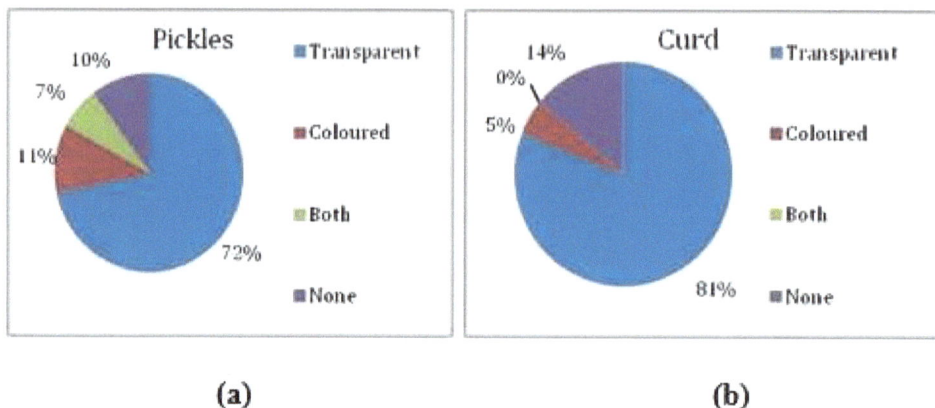

Fig. (22). Response to-while purchasing (a), pickles and (b) curd what kind of plastic bags do you use?.

Fig. (**22**) indicates that though most of the people used transparent plastic bags for pickles and curd, still some continued to use coloured plastic bags and were unaware of the leaching of colourants into food.

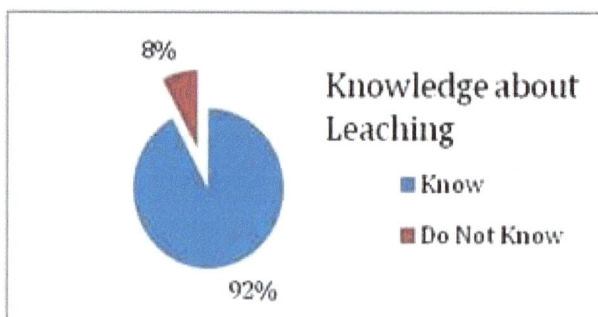

Fig. (23). Response to- Do you know that some chemicals leach out from plastics into stored food?.

Fig. (**23**) indicates that 92% of people were aware of the leaching of additives from plastics into food but continued using plastics to carry pickles and curd. Thus, there is an urgent need for converting knowledge/information into action.

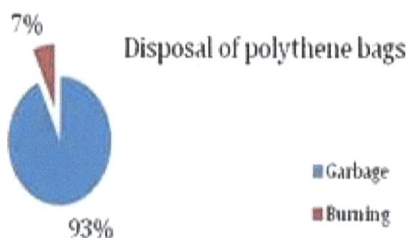

Fig. (24). Response to- how do you dispose of polythene bags (Burning/Garbage)?.

Fig. (**24**) indicates that still 7% of the polythene bags were being disposed of by burning. Even 7% is very harmful to the environment. Thus, there is a need for the generation of awareness about the harmful effects of plastic incineration.

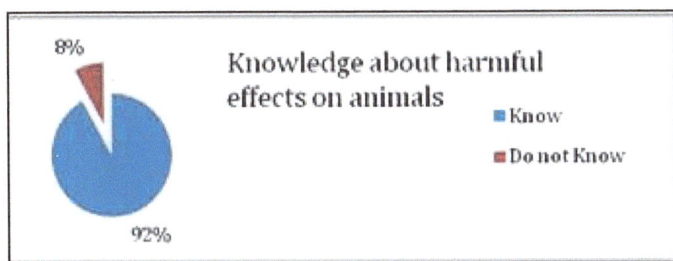

Fig. (25). Response to Have you observed animals accidentally eating plastic bags? Do you think it's harmful to them?.

Fig. (26). Response to- Do you practice segregation of garbage?.

Fig. **26** shows that only 36% of people practised segregation of garbage, thus even though the information is there, it has not been translated into action.

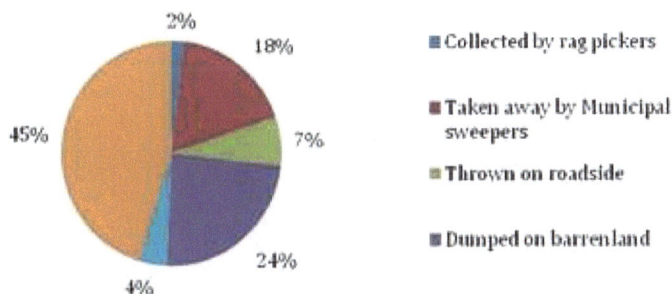

Fig. (27). Response to-What is the normal practice of disposal of biomedical waste in your locality?.

Fig. (**27**) highlights that 45% of people did not know about how biomedical wastes should be disposed of. Only 20% of waste that is collected by municipal sweepers and rag pickers was properly disposed of.

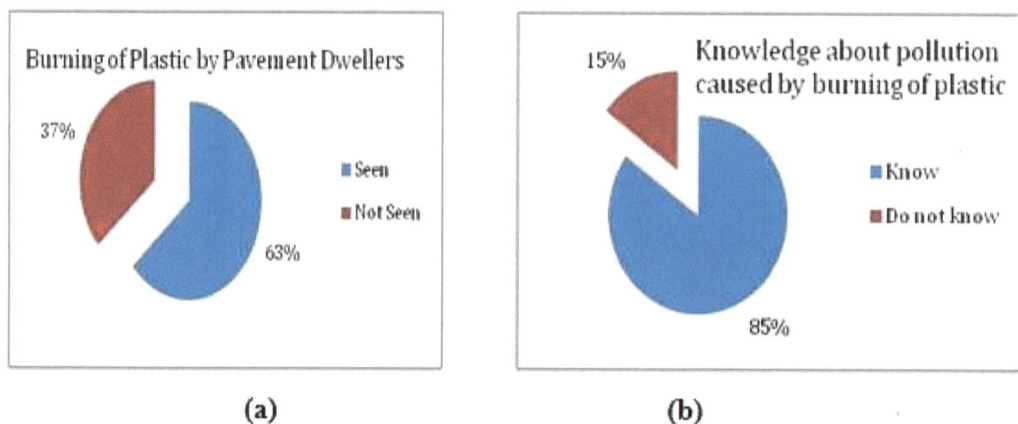

(a)   (b)

Fig. (28). Response to- **(a)**, Have you observed pavement dwellers burn plastic during winter? **(b)**, Should they be guided against inhaling gases evolved from it?

Fig. **(28)** points out that even a high level of awareness about the hazards of burning plastics has not helped in bringing about a positive change in this regard, in the external environment or locality.

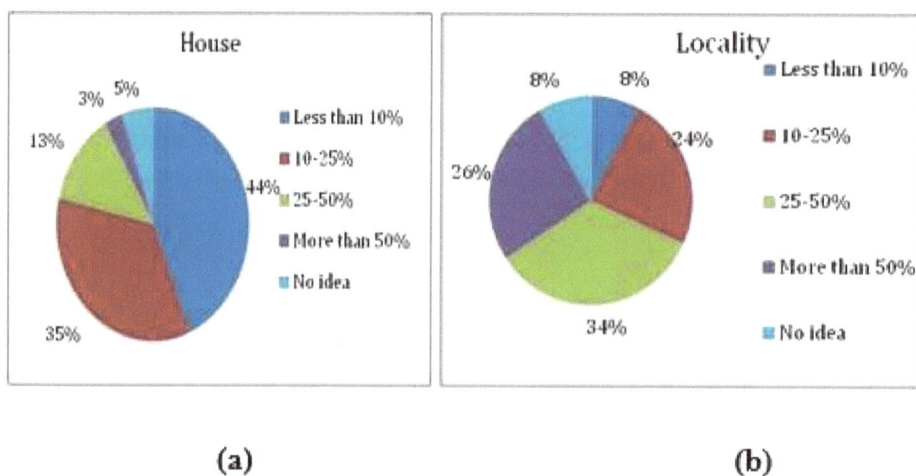

(a)   (b)

Fig. (29). Response to-What is the ratio (%) of plastic in the garbage in-(a) House, (b) Locality?.

Fig. **(29)** reflects the disparity and difference among people regarding their behaviour as responsible social beings. A total of 202 households (44%) considered that their garbage contains less than 10% of plastic waste, while only 36 households (8%) considered that their locality's garbage has less than 10% of plastic waste.

The survey shows that most people, though aware of leaching of additives and proper plastic waste management techniques, showed a disinclination to accept that they have a role to play in minimising and managing waste. Thus, activities centered around developing habits for judicious usage of plastics, segregation of garbage, and proper waste disposal are required.

AWARENESS BUILDING

A total of 23 undergraduate students were trained about safety evaluation, judicious usage, segregation, and waste disposal techniques for plastics. They developed a puppet show to sensitise the community interestingly. Awareness programmes were conducted in six city schools and three slum areas of Lucknow.

CONCLUDING REMARKS

Though in India the per capita consumption of plastic is 11kg, it is low as compared to the global average of 28 kg. However, we end up generating 5.6 million tonnes of plastic waste annually. Out of this, 60% is recycled, again a much better performance than the global average of 20%. The major challenge is that more than 60% of the recycling industry falls in the unorganised sector. The small recyclers lack infrastructure for proper treatment and cleaning of plastic waste, leading to the release of effluents, dust, and debris into the environment. Moreover, due to the lack of state-of-the-art testing facilities, fewer or no quality checks are performed before the finished products are sold in the market.

In the present study, the leachates of 30% samples displayed global migration above permissible limits and in leachates of some sample's cadmium, chromium and lead could be detected above the permissible limit. Children daily carry these tiffin boxes and bottles to schools. This long-term exposure to additive migration could be detrimental to their health.

The way ahead can be to set up plastic parks in various cities wherein a cluster of small-scale recycling units can be set up with the common state of the art testing facilities and effluent treatment plant so that every plastic commodity available in the market subscribes to the laid down BIS specifications. There is a need for more stringent waste management rules and their strict enforcement.

Everyone needs to play their part in following the strategy of 'Reduce, Reuse and Recycle' for plastic waste management. This should be done in the same priority order. The biggest irony is that of the total plastic generated annually worldwide, and more than 50% is thrown away within a few minutes after its first use. Just for one-time use, we are creating something which will be dumped in our environment for years. So, the most important step is to put a complete ban on single-use plastic. More than 60 countries have already done this. The government in India is also aiming to limit the consumption of single-use plastic and eventually eliminate it by 2022, but more stringent rules and their enforcement is needed.

The survey points out that awareness about the advantages of waste segregation exists and most of us understand the gravity of the situation, but still do not bring waste segregation into practice. The success of recycling depends completely on how well the waste has been segregated at the source. Though waste management is being practised in some cities in India, it is yet to be followed in larger parts of the country. We need to use plastics judiciously; though it is good to reuse them, the plastic articles should be used strictly for what they have been prescribed. Single-use plastic bottles should not be used over time; hot liquids should not be stored in these bottles. Food should not be heated in the microwave in plastic containers. High-fat content foods and very acidic or very basic foods should not be stored in plastic containers. The best strategy to reduce toxicity due to migration of additives seems to be to reduce exposure to additives, increase the rate of their elimination from our body and build up the antioxidants level in our body to mitigate oxidative stress triggered by additive overload.

CONSENT FOR PUBLICATION

Not Applicable.

CONFLICT OF INTEREST

The author confirms that this chapter contents have no conflict of interest.

ACKNOWLEDGEMENT

Declared none.

REFERENCES

[1]   www.plasticseurope.org

[2]   Mohanty S. Recycling of Plastics in Indian Perspective. UNIDO 2018.

[3]   Srivastava SP, Saxena AK, Seth PK. Safety evaluation of some of the commonly used plastic materials in India. Indian J Environ Health 1984; 26(4): 346-54.

[4]   Bhunia K, Sablani SS, Tang J, Rasco B. Migration of chemical compounds from packaging polymers during microwave, conventional heat treatment, and storage. Compr Rev Food Sci Food Saf 2013; 12(5): 523-45.
        [http://dx.doi.org/10.1111/1541-4337.12028] [PMID: 33412668]

[5]   Khan S, Khan A. Toxic heavy metal contamination in locally made plastic food container. Int J Sci Eng Res 2015; 6.

[6]   Genuis SJ, Beesoon S, Lobo RA, Birkholz D. Human elimination of phthalate compounds: blood,

urine, and sweat (BUS) study. ScientificWorldJournal 2012; 2012615068
[http://dx.doi.org/10.1100/2012/615068] [PMID: 23213291]

[7]   Genuis SJ, Beesoon S, Birkholz D, Lobo RA. Human excretion of bisphenol A: blood, urine, and sweat (BUS) study. J Environ Public Health 2012; 2012185731
[http://dx.doi.org/10.1155/2012/185731] [PMID: 22253637]

[8]   Determination of Overall Immigration of Constituents of Plastic Materials and Articles Intended to Come in Contact with Food Stuff-Method of Analysis (9845). Bureau of Indian Standards 1998.

[9]   Tchounwou PB, Yedjou CG, Patlolla AK, Sutton DJ. Heavy metal toxicity and the environment.Molecular, Clinical and Environmental Toxicology; Luch, A. Basel 2012; Vol. 101: pp. 133-64.
[http://dx.doi.org/10.1007/978-3-7643-8340-4_6]

[10]   Rodrigues EG, Bellinger DC, Valeri L, *et al.* Neurodevelopmental outcomes among 2- to 3-year-old children in Bangladesh with elevated blood lead and exposure to arsenic and manganese in drinking water. Environ Health 2016; 15(1): 44.
[http://dx.doi.org/10.1186/s12940-016-0127-y] [PMID: 26968381]

[11]   https://www.who.int/news-room/fact-sheets/detail/lead-poisoningand-health

[12]   Mathew B. Lipid peroxidation: Mechanism, models and significance. Int J Curr Sci 2012; 3: 11-7.

[13]   Flora G, Gupta D, Tiwari A. Toxicity of lead: A review with recent updates. Interdiscip Toxicol 2012; 5(2): 47-58.
[http://dx.doi.org/10.2478/v10102-012-0009-2] [PMID: 23118587]

[14]   Castagnetto JM, Hennessy SW, Roberts VA, Getzoff ED, Tainer JA, Pique ME. MDB: The metalloprotein database and browser at the scripps research institute. Nucleic Acids Res 2002; 30(1): 379-82.
[http://dx.doi.org/10.1093/nar/30.1.379] [PMID: 11752342]

[15]   Zhang D, Gao M, Jin Q, Ni Y, Zhang J. Updated developments on molecular imaging and therapeutic strategies directed against necrosis. Acta Pharm Sin B 2019; 9(3): 455-68.
[http://dx.doi.org/10.1016/j.apsb.2019.02.002] [PMID: 31193829]

[16]   Flora SJS, Pachauri V. Chelation in metal intoxication. Int J Environ Res Public Health 2010; 7(7): 2745-88.
[http://dx.doi.org/10.3390/ijerph7072745] [PMID: 20717537]

[17]   Miller AL. Dimercaptosuccinic acid (DMSA), a non-toxic, water-soluble treatment for heavy metal toxicity. Altern Med Rev 1998; 3(3): 199-207.
[PMID: 9630737]

[18]   Andersen O. Principles and recent developments in chelation treatment of metal intoxication. Chem Rev 1999; 99(9): 2683-710.
[http://dx.doi.org/10.1021/cr980453a] [PMID: 11749497]

[19]   Seaton CL, Lasman J, Smith DR. The effects of CaNa(2)EDTA on brain lead mobilization in rodents determined using a stable lead isotope tracer. Toxicol Appl Pharmacol 1999; 159(3): 153-60.
[http://dx.doi.org/10.1006/taap.1999.8725] [PMID: 10486301]

[20]   Bradberry S, Vale A. A comparison of sodium calcium edetate (edetate calcium disodium) and succimer (DMSA) in the treatment of inorganic lead poisoning. Clin Toxicol (Phila) 2009; 47(9): 841-58.
[http://dx.doi.org/10.3109/15563650903321064] [PMID: 19852620]

[21]   Britton RS, Leicester KL, Bacon BR. Iron toxicity and chelation therapy. Int J Hematol 2002; 76(3): 219-28.
[http://dx.doi.org/10.1007/BF02982791] [PMID: 12416732]

[22]   Cao Y, Skaug MA, Andersen O, Aaseth J. Chelation therapy in intoxications with mercury, lead and

copper. J Trace Elem Med Biol 2015; 31: 188-92.
[http://dx.doi.org/10.1016/j.jtemb.2014.04.010] [PMID: 24894443]

[23]   Aaseth J, Skaug MA, Cao Y, Andersen O. Chelation in metal intoxication--Principles and paradigms. J Trace Elem Med Biol 2015; 31: 260-6.
[http://dx.doi.org/10.1016/j.jtemb.2014.10.001] [PMID: 25457281]

[24]   Flora SJS, Chouhan S, Kannan GM, Mittal M, Swarnkar H. Combined administration of taurine and monoisoamyl DMSA protects arsenic induced oxidative injury in rats. Oxid Med Cell Longev 2008; 1(1): 39-45.
[http://dx.doi.org/10.4161/oxim.1.1.6481] [PMID: 19794907]

[25]   Flora SJS, Bhadauria S, Pachauri V, Yadav A. Monoisoamyl 2, 3-dimercaptosuccinic acid (MiADMSA) demonstrates higher efficacy by oral route in reversing arsenic toxicity: a pharmacokinetic approach. Basic Clin Pharmacol Toxicol 2012; 110(5): 449-59.
[http://dx.doi.org/10.1111/j.1742-7843.2011.00836.x] [PMID: 22117535]

[26]   Saxena G, Flora SJS. Lead-induced oxidative stress and hematological alterations and their response to combined administration of calcium disodium EDTA with a thiol chelator in rats. J Biochem Mol Toxicol 2004; 18(4): 221-33.
[http://dx.doi.org/10.1002/jbt.20027] [PMID: 15452883]

[27]   Pande M, Flora SJS. Lead induced oxidative damage and its response to combined administration of alpha-lipoic acid and succimers in rats. Toxicology 2002; 177(2-3): 187-96.
[http://dx.doi.org/10.1016/S0300-483X(02)00223-8] [PMID: 12135622]

[28]   Serrano SE, Braun J, Trasande L, Dills R, Sathyanarayana S. Phthalates and diet: a review of the food monitoring and epidemiology data. Environ Health 2014; 13(1): 43.
[http://dx.doi.org/10.1186/1476-069X-13-43] [PMID: 24894065]

[29]   Sioen I, Fierens T, Van Holderbeke M, *et al.* Phthalates dietary exposure and food sources for Belgian preschool children and adults. Environ Int 2012; 48: 102-8.
[http://dx.doi.org/10.1016/j.envint.2012.07.004] [PMID: 22885666]

[30]   Nohynek GJ, Borgert CJ, Dietrich D, Rozman KK. Endocrine disruption: fact or urban legend? Toxicol Lett 2013; 223(3): 295-305.
[http://dx.doi.org/10.1016/j.toxlet.2013.10.022] [PMID: 24177261]

[31]   Kim JH, Park HY, Bae S, Lim Y-H, Hong Y-C. Diethylhexyl phthalates is associated with insulin resistance *via* oxidative stress in the elderly: a panel study. PLoS One 2013; 8(8)e71392
[http://dx.doi.org/10.1371/journal.pone.0071392] [PMID: 23977034]

[32]   Ye H, Ha M, Yang M, Yue P, Xie Z, Liu C. Di2-ethylhexyl phthalate disrupts thyroid hormone homeostasis through activating the Ras/Akt/TRHr pathway and inducing hepatic enzymes. Sci Rep 2017; 7: 40153.
[http://dx.doi.org/10.1038/srep40153] [PMID: 28065941]

[33]   Hatch EE, Nelson JW, Qureshi MM, *et al.* Association of urinary phthalate metabolite concentrations with body mass index and waist circumference: a cross-sectional study of NHANES data, 1999-2002. Environ Health 2008; 7: 27.
[http://dx.doi.org/10.1186/1476-069X-7-27] [PMID: 18522739]

[34]   Wang H, Zhou Y, Tang C, *et al.* Urinary phthalate metabolites are associated with body mass index and waist circumference in Chinese school children. PLoS One 2013; 8(2)e56800
[http://dx.doi.org/10.1371/journal.pone.0056800] [PMID: 23437242]

[35]   Rowdhwal SSS, Chen J. Toxic Effects of Di-2-ethylhexyl Phthalate: An Overview. BioMed Res Int 2018; 20181750368
[http://dx.doi.org/10.1155/2018/1750368] [PMID: 29682520]

[36]   Upson K, Sathyanarayana S, De Roos AJ, *et al.* Phthalates and risk of endometriosis. Environ Res 2013; 126: 91-7.

[http://dx.doi.org/10.1016/j.envres.2013.07.003] [PMID: 23890968]

[37]   Calafat AM, Ye X, Wong L-Y, Reidy JA, Needham LL. Exposure of the U.S. population to bisphenol A and 4-tertiary-octylphenol: 2003-2004. Environ Health Perspect 2008; 116(1): 39-44. [http://dx.doi.org/10.1289/ehp.10753] [PMID: 18197297]

[38]   Rubin BS, Bisphenol A. Bisphenol A: an endocrine disruptor with widespread exposure and multiple effects. J Steroid Biochem Mol Biol 2011; 127(1-2): 27-34. [http://dx.doi.org/10.1016/j.jsbmb.2011.05.002] [PMID: 21605673]

[39]   Konieczna A, Rutkowska A, Rachoń D. Health risk of exposure to Bisphenol A (BPA). Rocz Panstw Zakl Hig 2015; 66(1): 5-11. [PMID: 25813067]

[40]   Ma Y, Liu H, Wu J, *et al.* The adverse health effects of bisphenol A and related toxicity mechanisms. Environ Res 2019; 176108575 [http://dx.doi.org/10.1016/j.envres.2019.108575] [PMID: 31299621]

[41]   Indirect Food Additives. Federal Register. Food and Drug Administration 2012; Vol. 77.

[42]   Federal Register. Indirect Food Additives. Adhesives AndComponents of CoatingsFood and Drug Administration 2013; 78.

[43]   Rochester JR, Bolden AL. Bisphenol S and F: A systematic review and comparison of the hormonal activity of bisphenol a substitutes. Environ Health Perspect 2015; 123(7): 643-50. [http://dx.doi.org/10.1289/ehp.1408989] [PMID: 25775505]

[44]   Chouhan S, Yadav SK, Prakash J. Swati; Singh, S. P. Effect of bisphenol a on human health and its degradation by microorganismsa: A review. Ann Microbiol 2014; 64: 13-21. [http://dx.doi.org/10.1007/s13213-013-0649-2]

[45]   Lorber M, Schecter A, Paepke O, Shropshire W, Christensen K, Birnbaum L. Exposure assessment of adult intake of bisphenol A (BPA) with emphasis on canned food dietary exposures. Environ Int 2015; 77: 55-62. [http://dx.doi.org/10.1016/j.envint.2015.01.008] [PMID: 25645382]

[46]   Hormann AM, Vom Saal FS, Nagel SC, *et al.* Holding thermal receipt paper and eating food after using hand sanitizer results in high serum bioactive and urine total levels of bisphenol A (BPA). PLoS One 2014; 9(10)e110509 [http://dx.doi.org/10.1371/journal.pone.0110509] [PMID: 25337790]

[47]   Geens T, Goeyens L, Kannan K, Neels H, Covaci A. Levels of bisphenol-A in thermal paper receipts from Belgium and estimation of human exposure. Sci Total Environ 2012; 435-436: 30-3. [http://dx.doi.org/10.1016/j.scitotenv.2012.07.001] [PMID: 22846760]

[48]   Pottenger LH, Domoradzki JY, Markham DA, Hansen SC, Cagen SZ, Waechter JM Jr. The relative bioavailability and metabolism of bisphenol A in rats is dependent upon the route of administration. Toxicol Sci 2000; 54(1): 3-18. [http://dx.doi.org/10.1093/toxsci/54.1.3] [PMID: 10746927]

[49]   Bernier MR, Vandenberg LN. Handling of thermal paper: Implications for dermal exposure to bisphenol A and its alternatives. PLoS One 2017; 12(6)e0178449 [http://dx.doi.org/10.1371/journal.pone.0178449] [PMID: 28570582]

[50]   Bach C, Dauchy X, Severin I, Munoz J-F, Etienne S, Chagnon M-C. Effect of sunlight exposure on the release of intentionally and/or non-intentionally added substances from polyethylene terephthalate (PET) bottles into water: chemical analysis and *in vitro* toxicity. Food Chem 2014; 162: 63-71. [http://dx.doi.org/10.1016/j.foodchem.2014.04.020] [PMID: 24874358]

[51]   Kieffer DA, Martin RJ, Adams SH. Impact of dietary fibers on nutrient management and detoxification organs: Gut, liver, and kidneys. Adv Nutr 2016; 7(6): 1111-21. [http://dx.doi.org/10.3945/an.116.013219] [PMID: 28140328]

[52]   Rahman MS, Kang K-H, Arifuzzaman S, *et al.* Effect of antioxidants on BPA-induced stress on sperm function in a mouse model. Sci Rep 2019; 9(1): 10584.
[http://dx.doi.org/10.1038/s41598-019-47158-9] [PMID: 31332285]

[53]   Abd El-Fattah AA, Fahim AT, Sadik NAH, Ali BM. Resveratrol and curcumin ameliorate di-(--ethylhexyl) phthalate induced testicular injury in rats. Gen Comp Endocrinol 2016; 225: 45-54.
[http://dx.doi.org/10.1016/j.ygcen.2015.09.006] [PMID: 26361869]

[54]   Abdel-Rahman HG, Abdelrazek HMA, Zeidan DW, Mohamed RM, Abdelazim AM. Lycopene: Hepatoprotective and antioxidant effects toward bisphenol a-induced toxicity in female wistar rats. Oxid Med Cell Longev 2018; 20185167524
[http://dx.doi.org/10.1155/2018/5167524] [PMID: 30147835]

[55]   Güleş Ö, Kum Ş, Yıldız M, *et al.* Protective effect of coenzyme Q10 against bisphenol-A-induced toxicity in the rat testes. Toxicol Ind Health 2019; 35(7): 466-81.
[http://dx.doi.org/10.1177/0748233719862475] [PMID: 31364507]

[56]   Hornos Carneiro MF, Shin N, Karthikraj R, Barbosa F Jr, Kannan K, Colaiácovo MP. Antioxidant CoQ10 restores fertility by rescuing bisphenol A-Induced oxidative DNA damage in the *Caenorhabditis elegans* germline. Genetics 2020; 214(2): 381-95.
[http://dx.doi.org/10.1534/genetics.119.302939] [PMID: 31852725]

CHAPTER 7

Natural Compounds with Anticancer Therapeutic Potential for Combating Ecotoxic Carcinogens

Anamika Mishra[1] and **Nidhi Mishra**[1,*]

[1] *Indian Institute of Information Technology Allahabad, Prayagraj, India*

Abstract: Cancer is a disease characterized by uncontrolled proliferation of cells, and it is caused due to the complex interaction of many cancer-causing factors that change the normal functioning of some genes; these factors may be internal causing mutation, or they may be ecotoxic carcinogens. Ecotoxic carcinogens are the agents that are present in our environment and exposure to them can increase the risk of cancer. These include aflatoxins, arsenic, asbestos, coke-oven emissions, tobacco smoke, wood dust, and indoor emissions from the household combustion of coal, *etc.* Nature has provided us with an enormous source of natural compounds which have application in various fields such as medicine, cosmetics industry, food, and nutrition, *etc.* Nature and the natural compounds are serving as a boon to mankind. The medicinal application of natural compounds is one of the most prominent applications of plant products. Plant products have been playing a very important role in the treatment of cancer, as many anticancer drugs have been developed from plant products, such as vinca alkaloids (vinblastine, vincristine, and vindesine), the epipodophyllotoxins (etoposide and teniposide), the taxanes (paclitaxel and docetaxel) and the camptothecin derivatives (camptothecin and irinotecan), *etc.* The plant-derived anti-cancer drugs have benefits such as easy availability, cost-effectiveness, and fewer side effects. This book chapter will emphasize the various ecotoxic carcinogens and numerous plant-derived anti-cancer drugs, with their mechanism of action.

Keywords: Amphiboles, Anthropogenic, Apoptosis, Chromosomal aberrations, DNA adducts, DNA methylation, Genotoxic, Nitropolyaromatic, Oncogenes.

INTRODUCTION

Cancer is a serious health issue across the globe and one of the major reasons for deaths reported globally. According to the World Health Organisation (WHO), cancer is defined as uncontrolled growth and spread of cells, which can affect any

* **Corresponding author Nidhi Mishra:** Indian Institute of Information Technology, Prayagraj, India; E-mail: nidhimishra@iiita.ac.in

Tahmeena Khan, Abdul Rahman Khan, Saman Raza, Iqbal Azad and Alfred J. Lawrence (Eds.)

part of the body. Apart from this, cancer is characterized by the invasion of the surrounding tissue and tends to metastasize the distant sites too [1]. The cancerous cells differ from the normal cells in the following manner:

1. Cancer cells show uncontrolled proliferation, and this is accomplished in many ways, such as by the generation of autocrine stimulation, not responding to the inhibitory signals and apoptotic signals, lacking contact inhibition. Normal cells are genetically stable, as they have machinery that works to rectify the mutations that occur during cellular processes, while the cancerous cells have mutations and chromosomal aberrations as well, which makes the cancerous cells genetically unstable. Cancer cells are immortal, while normal cells show cellular senescence *i.e.*, they die after performing a particular number of cell divisions, the cancerous cells keep on dividing. This happens due to the presence of telomerase enzyme in cancerous cells which lacks in the case of normal cells. The telomerase enzyme in cancer cells helps to maintain the length of telomeres.

2. Cancer cells have the property of tissue invasion and metastasis which is not the property of normal cells [2, 3]. Our body is made up of billions of cells grouped to form tissues. The tissues are grouped to form organs; as different organs of our body are involved in doing different activities, so the cells of different organs differ from each other as well. Depending on the cell from where cancer originates cancer is divided into 5 categories [3, 4].

• Carcinoma: is the cancer of epithelial tissue, which forms the covering and lining of body organs and the body cavity. Carcinoma is further classified as adenocarcinoma, squamous cell carcinoma, basal cell carcinoma, and transitional cell carcinoma.

• Sarcoma: sarcoma is cancer that affects the connective tissue of the body and it is of two types, namely bone sarcoma and soft tissue sarcoma.

• Leukaemia: it is the cancer of blood-forming tissue.

• Lymphomas and myelomas: these are the cancer of the lymphatic system, which works to fight against infections.

• Brain and spinal cord cancer: cancer can also occur in the brain cells and cells of the spinal cord as well [4].

Cancer is caused due to alteration in gene function which occurs due to mutations and chromosomal aberrations [5]. These changes result in the cumulative effect of both internal and external factors, the internal factors being the cellular machinery plays a role in crucial cellular processes, like DNA replication, cell cycle progression, *etc.* while the external factors are the substances that act as carcinogens [6]. Carcinogens are compounds that have the potential to cause cancer in humans and model organisms [7]. Our environment also serves as a reservoir of many ecotoxic carcinogens. Due to the high levels of pollution, many contaminants are present in our environment that are genotoxic, which make them carcinogenic; even scientific reports have been published that state that the environment plays a role in the development of cancer [8]. However, our environment is not just the culprit; rather it has provided a way to fight against cancer as many anticancer drugs have been developed from plant-based sources [9].

ENVIRONMENT: A RESERVOIR OF ECOTOXIC CARCINOGENS AND SOURCE OF ANTICANCER DRUGS

Ecotoxic Carcinogens

The environment constitutes the surroundings in which an organism lives, it includes both biotic and abiotic components [10], and the organisms interact with different components of the environment. Likewise, humans interact with their surrounding environment to fulfill their specific needs that impose an adverse effect on environmental conditions [11]. The level of contaminants has increased in the environment and these contaminants affect human health as we are in direct contact with our environment [12]. Cancer is a disease characterized by uncontrolled proliferation of cells; causative agents of cancer include both genetic and ecotoxic carcinogens present in our environment. In other words, the development of cancer depends on the deposition of both genetic and epigenetic changes in the cell. The environment also influences the occurrence of cancer [13, 14]. The term ecotoxic carcinogen is used for pollutants and contaminants present in our environment having the potential to cause cancer [15]. The ecotoxic carcinogens include arsenic, asbestos, benzene, cadmium, smoke, gasoline, hair dye, nickel, radon, *etc* [13, 16]; some of them are explained below:

Aflatoxins

Aflatoxins are the secondary metabolites obtained from polyketides produced by different species of *Aspergillus* (*Aspergillus flavus, A. parasiticus*, and *A. nomius)*. Being highly toxic, these fungi are the contaminants of several important cereal crops, such as wheat, walnut, corn, cotton, peanuts, tree nuts, *etc* [16, 17]. Aflatoxins are reported to cause several health issues in humans and animals [17]. Aflatoxins (Aflatoxin B1) are one of the ecotoxic carcinogens which cause hepatocellular carcinoma [18]. People get exposure to aflatoxins when they come in contact with the crops that are contaminated with the fungi *Aspergillus* or when they encounter the affected animals. The four more common types of aflatoxins produced by different species of *Aspergillus* (*Aspergillus flavus, A. parasiticus,* and *A. Nomius*) are Aflatoxin B1 (Fig. **1**), B2, G1, and G2. Among them, B1 is the most potent carcinogen and is commonly found in contaminated crops [19].

In their original forms, Aflatoxins are nontoxic and are converted to mutagenic and carcinogenic form by the action of hepatic cytochrome p450 [18, 19]. Aflatoxins promote tumours by activating proto-oncogenes and causing mutations in tumour suppressor genes, for example, they are known to report a mutation in the p53 gene. Apart from this, they work synergistically with the Hepatitis B virus and increase the chance of HCC; they also cause chromosomal aberrations [18].

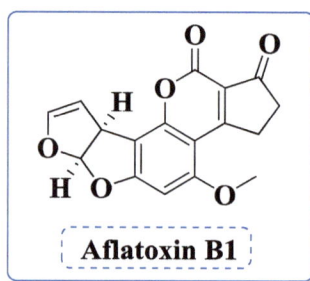

Fig. (1). Chemical structure of Aflatoxin B1.

Aristolochic Acids

Aristolochic acids (AA) (Fig. **2**) are naturally occurring compounds in the plants of genus *Aristolochia* and *Asarum* which are found worldwide, and are used as a herbal medicine for the treatment of several diseases, such as cold, headache, snake bites inflammatory diseases, *etc.* Chemically, it is a nitropolyaromatic compound [20, 21]. Aristolochic acids are reported to cause nephropathy, which is called aristolochic acids induced nephropathy (AAN), and upper tract urothelial carcinoma (UTUC) [20, 21]. The major components of AA are AA1 and AA2. Both are nitrophenanthrene carboxylic acids and mutagenic as they can form DNA adducts; after the activation by some enzymes their activation involves a reduction of the nitro group [20]. AA had been experimentally proved to be carcinogenic in rodents but later it was observed that they are carcinogenic to humans as well [21]. They show mutagenicity by reacting with DNA to form the aristolactam DNA (AL-DNA) adducts, they bind with purine bases to form dG-AL, and dA-AL, and these adducts act as mutagens by blocking the process of DNA replication; apart from this, dA-AL induced by AA1 causes A: T transversion in the non-transcribe segment of TP53 gene [20].

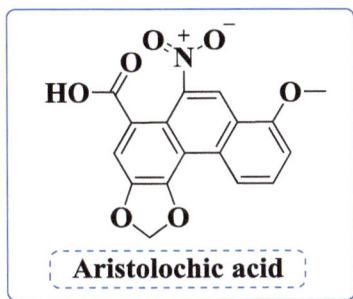

Fig. (2). Chemical structure of Aristolochic acid.

Arsenic

Arsenic is present in the environment; rocks, soil, and water serve as the reservoir for arsenic and it is also present in plants and animals. People get exposed to arsenic from agriculture and industrial sources. In the environment, arsenic is present in two forms: a less toxic organic form and a more toxic inorganic form [16]. It enters the environment throughdifferent ways, such as by the use of pesticides containing arsenic, processing of ores and minerals, present in tobacco, and even in some food materials, and also as a contaminant in water [22]. Long-term consumption of arsenic poses serious health issues that include skin lesions, hypertension, ischemia, some endemic peripheral vascular disorders, diabetes, severe arteriosclerosis, neuropathies, and significantly, many types of cancers (mainly skin and lung cancer) [22, 23].

Arsenic works as a carcinogen by affecting the normal cellular processes; this includes methylation of arsenic itself. The methylated arsenic acts as an activated, potential carcinogen that affects normal cellular processes, such as gene transcription, and serving as an inhibitor for enzymes. S-adenosylmethionine (SAM) is the donor of a methyl group for arsenic, this is not only the donor of a methyl group for arsenic rather it serves as the donor of a methyl group for other important cellular processes, such as in the case of DNA methylation, histone modification by methylation and regulation of miRNA expression by methylation. So, the methylation of arsenic by SAM affects the normal pool of methyl group and imposes pressure on the rest of the important cellular processes. Arsenic exposure also causes oxidative stress, which is characterized by the generation of Reactive Oxygen Species (ROS); the ROS causes instability in the cells in many ways, such as by causing DNA damage or targeting signalling pathways, *etc*. It is because of these properties that arsenic is a carcinogen [23 - 25].

Asbestos

Asbestos is the term used for naturally occurring mineral fibres that are carcinogenic. They have the property of heat resistance and corrosion resistance, which has made them suitable materials to use for many industrial purposes *e.g.*, insulation and fireproofing materials, automotive brakes, and wallboard materials. People get exposed to asbestos by inhaling or swallowing from the environment; people who are working in the industries that work with asbestos are more susceptible to get affected by it [16, 26, 27]. Asbestos is present in two forms, amphiboles that are straight and needle-like in shape and include many varieties of asbestos fibres such as crocidolite (blue asbestos) (Fig. **3a**), amosite (brown asbestos), tremolite, anthophyllite, and actinolite, while the other one is serpentine which is spiral and includes only chrysotile (Fig. **3b**) [16, 27]. Both these fibres are carcinogenic. It has been reported that exposure to asbestos leads to the occurrence of lung cancer and mesothelioma. Apart from this, it is also evident that its exposure causes many other types of cancer, such as in the larynx, ovary, colorectum, pharynx, and stomach [27 - 29]. The dimensions of the fibres affect the carcinogenicity of asbestos in the case of mesothelioma [29], while it works in synergism with cigarette smoke to cause lung cancer [27, 29]. Asbestos acts as a carcinogen by inducing chromosomal breakage and by oxidative stress, as the chromosomal breakage affects the activation or inactivation of proto-oncogenes and tumour suppressor genes, while oxidative stress is associated with the generation of reactive oxygen species that affects the normal cellular processes [27, 29].

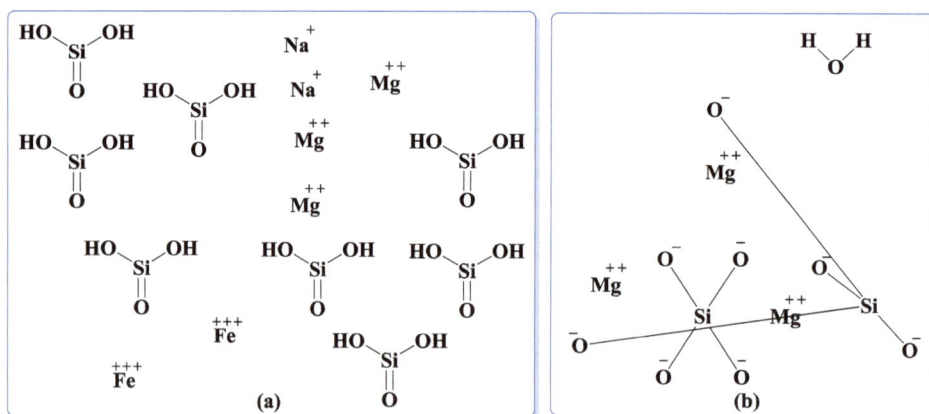

Fig. (3). Chemical structure of (a) crocidolite (b) chrysotile.

Benzene

Benzene is a commonly occurring air pollutant that enters the environment through various human activities. It is a component of gasoline, emission from vehicles and industries, tobacco smoke, and was used as a solvent in chemical industries. The people who work in industries where benzene is used get exposed to it, while others get exposed to it through the benzene present in the environment or by using the products that contain benzene. Benzene is reported as carcinogenic as it causes leukaemia *e.g.*, acute myeloid leukaemia, acute non-lymphocytic leukaemia, multiple myeloma, *etc.* Benzene is carcinogenic as it induces oxidative stress that causes genetic instability and genotoxicity and hence causes DNA damage, gene mutation, and chromosomal aberrations; it is also an inhibitor of topoisomerase II, which is an important enzyme that plays a role during DNA replication, responsible for formation DNA adducts [30 - 32].

Beryllium

Beryllium is a metal whose physicochemical properties make it a suitable material to produce various goods, such as nuclear weapons, dental prostheses, golf clubs, *etc.* The International Agency for Research on Cancer (IARC) has considered it as a group 1 carcinogen [33]. Its carcinogenicity has been tested against various cancer cell lines and in many animal models [33, 34]. It shows genotoxicity and cell transforming ability in different cell lines and it also binds to nucleoproteins, to inhibit certain enzymes that are required for the processes of DNA replication [33 - 35].

Cadmium

Cadmium is present in our environment and people get exposed to cadmium by inhalation of dust and fumes and also by incidental ingestion of dust from hands contaminated with cigarettes smoke, and foods while making and eating contaminated food. It has been reported as a carcinogen and primarily causes lung cancer, while several reports suggest that it plays a role in the development of prostate, kidney, and renal cancer as well. The carcinogenicity of cadmium occurs due to its ability to cause DNA damage and alteration in the expression of genes [16, 36].

Chromium

Chromium is a naturally occurring heavy metal, commonly used in industrial processes [37], which can cause severe health issues in humans. Chromium causes cancer of the lungs, paranasal sinuses, and nasal cavity [38]. The use of chromium has grown rapidly in industries, resulting in major anthropogenic environmental pollution. These industries include chromate or chromium production and plating, leather-tanning industries, commercial and industrial fuel (natural gas, coal, and oil), *etc.*, which are the largest contributors to chromium level. Chromium exists in soil, water, food, and air. Chromium occurs in two main valence states *i.e.*, Cr

(VI) and Cr (III). Cr (VI), which is extensively used in industries, is a human carcinogen; it enters the cells through sulphate/phosphate anion system. After entering the cell, it is reduced to the lower-valence intermediates Cr (V), Cr (IV), or Cr (III), *via* cellular reductants (glutathione and ascorbate). These intermediates may directly or indirectly damage DNA or result in DNA-protein cross-links. Cr (III) compounds are 100 times less toxic than Cr (VI) and are used as micronutrients and dietary supplements. Cr (III) can accumulate around cells because it cannot pass easily through the cell membrane and it disrupts the cellular function and integrity, and finally damage the DNA [39].

Cigarette Smoke

Cigarette smoking is a major risk factor for lung cancer as well as many other cancers, such as oesophageal, oral, laryngeal, pancreatic, bladder cancer, *etc.* Cigarette smoking is the main cause of cancer deaths in the world. Cigarette smoking is indirectly related to other risk factors, such as hypertension and hypercholesterolemia, which contribute to the development and promotion of the atherosclerotic process [40]. It is estimated that p^{53} is frequently the mutated gene in tobacco-related cancer and the mutation rate is higher in smokers than non-smokers. The mutation leads to G to T transversion and is seen in 30% of smokers and only 12% of non-smokers suffering from lung cancer [41].

Ionizing radiation

Ionizing radiations (IR) are high-energy radiations, such as x-rays and gamma-rays. They impose an adverse effect on human health in many ways. People get exposed to IR by both natural sources and anthropogenic sources. The natural source of IR is radioactive materials present in the soil, water, and air, while the anthropogenic sources are nuclear reactors, medical devices, *etc.* IR is carcinogenic as it induces DNA damage which leads to the development of cancer [42 - 44].

Apart from the above mentioned environmental carcinogens, there are many other factors present in our environment that are responsible for inducing cancer, for example, cancer-causing viruses like Hepatic virus-B and C which causes liver cancer, Burkitt's virus which induces cancer of lymph node, HPV/Herpes virus which causes cancer of cervix and skin, bacteria like Helicobacter pylori which induces stomach cancer, and many other chemicals, such as soot, vinyl chloride, nickel, formaldehyde, hair dyes, soot, *etc.* that enter the environment by different ways [13].

Environment for the Treatment of Cancer

The environment constitutes everything in our surroundings, including the living biotic components and nonliving abiotic components, and we have established a dynamic relationship with our surroundings. We get many beneficial things from our environment to fulfill our needs. One of these needs is to find medicines to fight against various diseases; plants that are one of the most important components of our environment serve as a source of medicine since time immemorial [45]. A wide diversity of flora exists on our planet that not only differs in physical appearance but chemical composition as well. So, the diverse groups of plants serve as a huge reservoir of different phytochemicals that are used by pharmaceutical industries to develop various drugs. Many anticancer drugs have been developed from plant sources; plant-derived anticancer drugs are a major component of available chemotherapy against cancer [46].

Vinca Alkaloids

Vinca alkaloids (VA) are the plant-derived anticancer drugs obtained from *Catharanthus roseus* (*C. roseus*), which is a member of the Apocynaceae family. They are currently in use for the treatment of several types of cancer like breast, liver, leukaemia, testes, and lung cancer. The four main types of VA include vinblastine (Fig. **4a**), vincristine (Fig. **4b**), vindesine (Fig. **4c**), and vinorelbine (Fig. **4d**). Their mechanism of action involves binding to tubulin heterodimers known as vinca-binding site, which leads to disruption of microtubule function, or they cease the cell cycle by arresting the cell in metaphase. Many semi-synthetic derivatives of vinca alkaloids, which include vinorelbine, vindesine, vinfosiltine, and vinorelbine, are also used as anticancer drugs. These VA derivatives are used alone or in combination with other plant-derived anticancer drugs, to fight against many cancers [46 - 48].

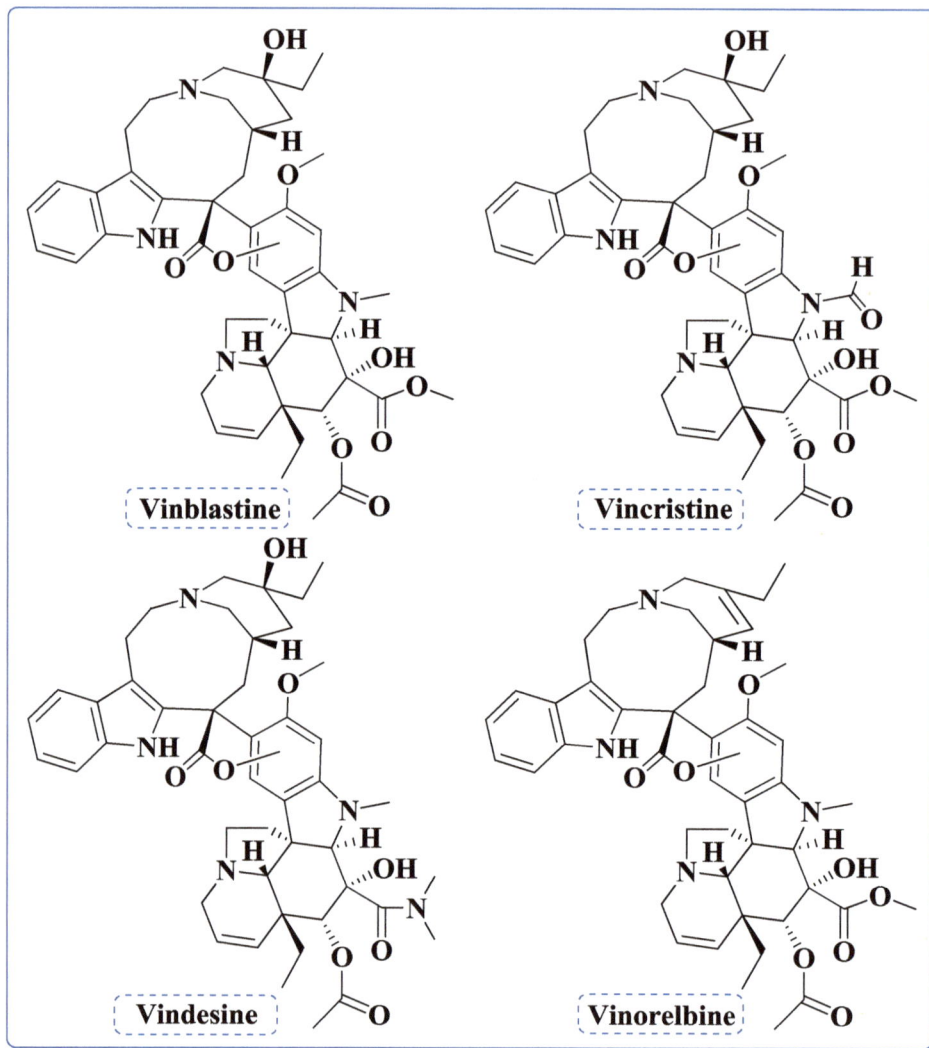

Fig. (4). Chemical structure of Vinca alkaloids, (a) Vinblastine, (b) Vincristine, (c) Vindesine and (d) Vinorelbine.

Taxanes

Taxanes are another very important group of anticancer drugs obtained from plants. They act by binding to microtubules that have a key role in cell division. The two effective taxanes are paclitaxel, which is isolated from the bark of the Pacific yew, *Taxus brevifolia*, and docetaxel, which is a semisynthetic analogue synthesized from DAB (10-deacetylbaccatin III) isolated from the leaves of the

European yew, *Taxus baccata.* Docetaxel (Fig. **5a**) and paclitaxel (Fig. **5b**) are strong anticancer agents in terms of their efficacy on different molecular targets. Paclitaxel is used to cure a wide range of cancers like ovarian, breast, and lung cancer, while docetaxel is used in breast, pancreas, prostate, and lung cancer therapies. The binding of paclitaxel with β-tubulin in the lumen of microtubules leads to a decrease in microtubule dynamics and halts the cell cycle at the M phase. The primary mechanism of taxanes is to induce microtubule stabilization, apoptotic cell death, and mitotic arrest. Analogues of paclitaxel that are currently undergoing clinical trials include larotaxel, milataxel, ortataxel, and tesetaxel [46, 47, 49].

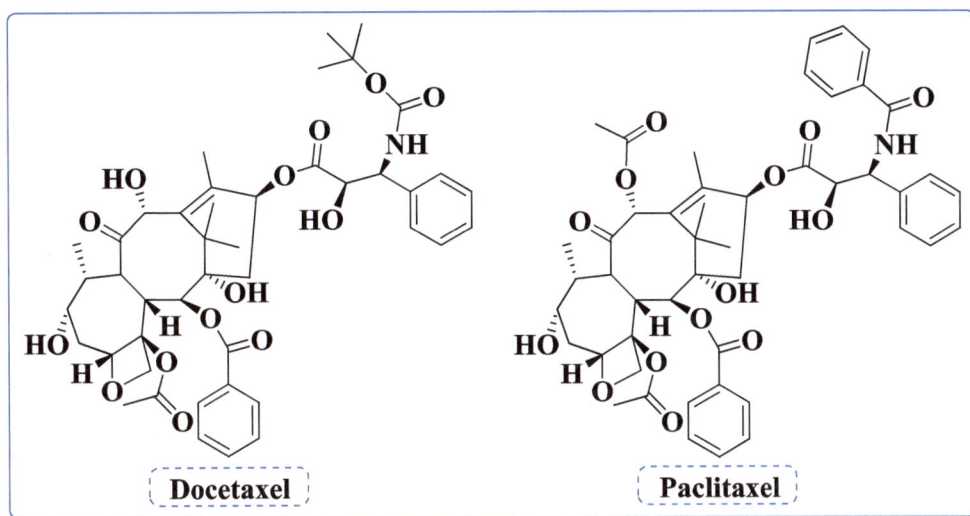

Fig. (5). Chemical structure of (a) docetaxel and (b) paclitaxel.

Podophyllotoxin

Podophyllotoxin (Fig. **6**) is derived from *Podophyllum peltatum*, also known as the American mayapple, and *Podophyllum emodii*. It has been used for the treatment of skin cancers and warts. The two most effective podophyllotoxin derivatives are etoposide and teniposide, which work against cancer. The podophyllotoxin binds to tubulin, while its derivatives *i.e.*, etoposide and teniposide inhibit topoisomerase II, inducing topoisomerase II-mediated DNA cleavage. Etoposide and teniposide are used for the treatment of lymphomas and bronchial and testicular cancers [46, 50].

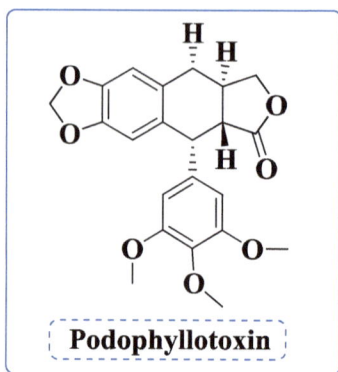

Podophyllotoxin

Fig. (6). Chemical structure of Podophyllotoxin.

Camptothecin Derivatives

Camptothecin is a plant-derived anti-cancer agent, that works by inhibiting topoisomerase I, thereby preventing DNA relegation and causes DNA damage, leading to apoptosis. It was first isolated from *Camptotheca* acuminate. Many semi-synthetic drugs have been developed from camptothecin, such as topotecan (Fig. **7a**), irinotecan (Fig. **7b**), and belotecan (Fig.**7c**). Irinotecan is used to treat colorectal cancer while topotecan is used to treat ovarian and lung cancer [46, 47, 51].

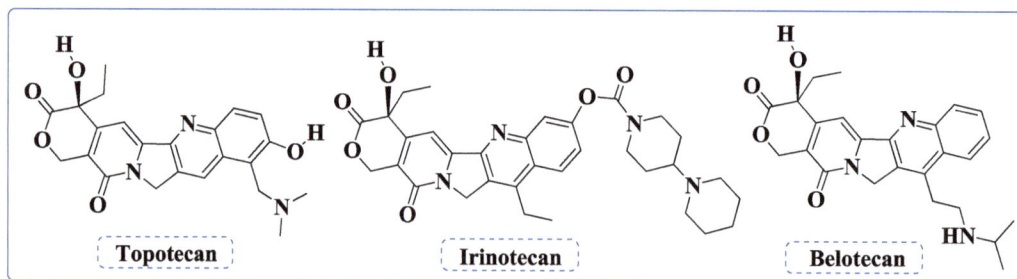

Topotecan   **Irinotecan**   **Belotecan**

Fig. (7). Chemical structure of Camptothecin derivatives (a) topotecan, (b) irinotecan, and (c) belotecan.

Colchicine

Colchicine (Fig. **8**) is a naturally occurring alkaloid, obtained from *Colchicum autumnale* (Colchicaceae), which is used for the treatment of diseases like crystal arthritis, cirrhosis, gout, *etc.* The ability of colchicines to bind with tubulin leads to stabilization of microtubule formation, leading to cell cycle arrest that ultimately causes cell death (apoptosis). This property of colchicine has made it a promising candidate to be used as an anticancer agent, but its activity is not specific for cancer cells, rather it targets every rapidly dividing normal cell and ceases the cell cycle. Colchicinamide and, Deacetylcolchicine are colchicine-derived semi-synthetic drugs, with less toxicity, that are used for the treatment of cancers, such as colorectal cancer (HCT-116), chronic granulocytic leukaemia, melanoma, central nervous system cancer, breast cancer, and gastric cancer [47, 52, 53].

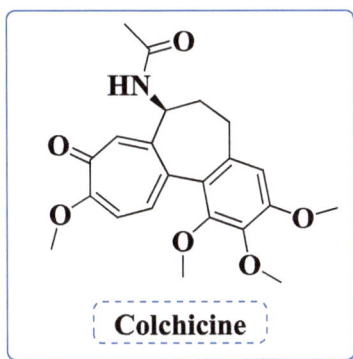

Fig. (8). Chemical structure of Colchicine.

Combretastatin

Combretastatin (Fig. **9**) is a member of stilbenes and was first obtained from *Combretum caffrum*. It is reported to inhibit angiogenesis by targeting endothelial cells; it also targets tubulin and actin cytoskeleton, causing changes in the structure of immature endothelial cells. In this way, it is responsible for depriving the tumour of essential nutrients [46, 47]. Modifications have been incorporated to develop many Combretastatin derivatives that have better activity and fewer side effects [54]. Combretastatin and its derivatives are used in the treatment of leukaemia, lung cancer, and colon cancer [47].

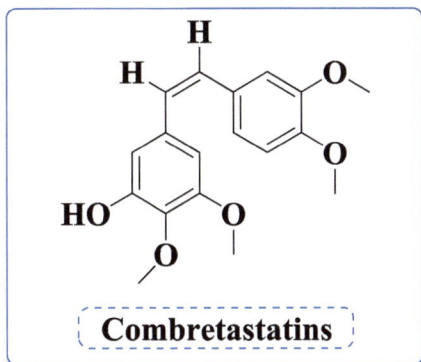

Fig. (9). Chemical structure of Combretastatin.

The list of plant-derived anticancer agents is quite long, as apart from the above-mentioned agents, many others are also an important component of chemotherapy; the examples are triterpenoid acids, capsaicin, flavonoids, gingerol, lycopene, *etc*. The drawbacks associated with the available chemotherapeutic agents, which include issues such as resistance, non-selectivity, rapid clearance, and bioavailability, have created the gap to search for better and effective anticancer agents. This is the reason that researchers are exploring various plant extracts, hoping to find a cure for cancer. Some other plant-derived anticancer agents that have been studied against various types of cancer are Geniposide and its derivatives (isolated from *Gardenia jasminoides* Ellis, Rubiaceae), Homoharringtonine (obtained from the bark of various cephalotaxus species), which is known as an anti-leukaemia drug, Roscovitine (derived from *Raphanus sativus,* Brassicaceae), Elipticine (isolated from *Ochrosia elliptica*), Salvicine (derived from *Salvia pronitis*), *etc* [55]. Plants are not the sole component of the environment serving for the development of anticancer agents. There are other sources as well, that are present in our environment, serving as a source for the development of anticancer drugs, such as marine organisms, bacteria, and fungi. Marine organisms are present in the deep sea and many compounds have been obtained from them that have anticancer properties. Examples are Cytarabine, which is an anti-leukemic agent, Clofarabine, which is a purine nucleoside analogue and is used in the treatment of leukaemia, Trabectedin, which works against soft-tissue sarcoma and Halichondrin B, which acts as a tubulin destabilizing agent. Many antibiotics, such as Rapamycin, Carfilzomib, and Midostaurin, are used in the treatment of cancer and are obtained from different fungi [46]. In this way, the environment plays a very significant role in the treatment of cancer.

Table 1. Anticancer agents with their roles in the prevention of cancer.

S.No.	Name of the Anticancer Agent	Activity	Reference
1	Geniposide and its derivatives	Cytotoxic activity against various cancer cell lines, affect cancer metastasis	[55, 56]
2	Homoharrigtonine	Anti-leukaemia drug	[55, 57]
4	Elipticine	Causes cell cycle arrest	[55, 58]
5	Roscovitine	Causes cell cycle arrest and apoptosis	[55, 59]
6	Maytansin	Targets microtubule assembly formation	[55]
8	Thapsigargin	Inhibitor of Ca^{2+}-ATPase, promotes apoptosis	[55, 60]

CONCLUDING REMARKS

Cancer is a disease whose occurrence depends on both genetic and epigenetic factors. The default in the gene system of the individuals is stimulated by internal factors, which include the error in the normal cellular processes, along with external factors that also play pivotal roles in the development of cancer. The external factors are the carcinogens present in our environment. Environmental carcinogens are majorly related to occupational cancer, a term which is used for cancer whose occurrence is associated with industries; the people working in such industries get exposed to the carcinogens and are more prone to develop cancer. Apart from being a source of carcinogens, the environment also acts as a reservoir for the development of many anticancer drugs.

CONSENT FOR PUBLICATION

Not Applicable.

CONFLICT OF INTEREST

The author confirms that this chapter contents have no conflict of interest.

ACKNOWLEDGEMENT

Declared none.

REFERENCES

[1]   World Health Organization. https://www.who.int/topics/cancer/en/

[2]   Rosenberg LE, Rosenberg DD. Human Genes and Genomes: Science, Health, Society. 1st ed., London; Waltham, MA: Elsevier Academic Press 2012.

[3]   Cooper GM. The Cell: A Molecular Approach. 2nd ed., Sunderland, MA: Sinauer Associates 2000.

[4]   https://www.cancerresearchuk.org/what-is-cancer/how-cancer-starts/types-of-cancer#main

[5]   Koeffler HP, McCormick F, Denny C. Molecular mechanisms of cancer. West J Med 1991; 155(5): 505-14.
        [PMID: 1815390]

[6]   Podolskiy DI, Gladyshev VN. Intrinsic *versus* extrinsic cancer risk factors and aging. Trends Mol Med 2016; 22(10): 833-4.
        [http://dx.doi.org/10.1016/j.molmed.2016.08.001] [PMID: 27544777]

[7]   Schrenk D. What is the meaning of 'A compound is carcinogenic'? Toxicol Rep 2018; 5: 504-11.
        [http://dx.doi.org/10.1016/j.toxrep.2018.04.002] [PMID: 29854622]

[8]   Blackadar CB. Historical review of the causes of cancer. World J Clin Oncol 2016; 7(1): 54-86.
        [http://dx.doi.org/10.5306/wjco.v7.i1.54] [PMID: 26862491]

[9]   Desai AG, Qazi GN, Ganju RK, *et al.* Medicinal plants and cancer chemoprevention. Curr Drug Metab 2008; 9(7): 581-91.
        [http://dx.doi.org/10.2174/138920008785821657] [PMID: 18781909]

[10] https://www.soas.ac.uk/courseunits/P120(P500).html

[11] Galvani AP, Bauch CT, Anand M, Singer BH, Levin SA. Human-environment interactions in population and ecosystem health. Proc Natl Acad Sci USA 2016; 113(51): 14502-6.
[http://dx.doi.org/10.1073/pnas.1618138113] [PMID: 27956616]

[12] Remoundou K, Koundouri P. Environmental effects on public health: an economic perspective. Int J Environ Res Public Health 2009; 6(8): 2160-78.
[http://dx.doi.org/10.3390/ijerph6082160] [PMID: 19742153]

[13] Parsa N. Environmental factors inducing human cancers. Iran J Public Health 2012; 41(11): 1-9.
[PMID: 23304670]

[14] Lewandowska AM, Rudzki M, Rudzki S, Lewandowski T, Laskowska B. Environmental risk factors for cancer - review paper. Ann Agric Environ Med 2019; 26(1): 1-7.
[http://dx.doi.org/10.26444/aaem/94299] [PMID: 30922021]

[15] Boffetta P, Nyberg F. Contribution of environmental factors to cancer risk. Br Med Bull 2003; 68: 71-94.
[http://dx.doi.org/10.1093/bmp/ldg023] [PMID: 14757710]

[16] https://www.cancer.gov/about-cancer/causes-prevention/risk/substances

[17] Kumar P, Mahato D K, Kamle M, Mohanta T K, Kang S G. A Global Concern for Food Safety, Human Health and Their Management. Front. Microbiol. ,2017, 07.

[18] Magnussen A, Parsi MA. Aflatoxins, hepatocellular carcinoma and public health. World J Gastroenterol 2013; 19(10): 1508-12.
[http://dx.doi.org/10.3748/wjg.v19.i10.1508] [PMID: 23539499]

[19] Kew MC. Aflatoxins as a cause of hepatocellular carcinoma. J Gastrointestin Liver Dis 2013; 22(3): 305-10.
[PMID: 24078988]

[20] Sidorenko VS, Attaluri S, Zaitseva I, *et al.* Bioactivation of the human carcinogen aristolochic acid. Carcinogenesis 2014; 35(8): 1814-22.
[http://dx.doi.org/10.1093/carcin/bgu095] [PMID: 24743514]

[21] Arlt VM, Stiborova M, Schmeiser HH. Aristolochic acid as a probable human cancer hazard in herbal remedies: a review. Mutagenesis 2002; 17(4): 265-77.
[http://dx.doi.org/10.1093/mutage/17.4.265] [PMID: 12110620]

[22] Pershagen G. The carcinogenicity of arsenic. Environ Health Perspect 1981; 40: 93-100.
[http://dx.doi.org/10.1289/ehp.814093] [PMID: 7023936]

[23] Martinez VD, Vucic EA, Becker-Santos DD, Gil L, Lam WL. Arsenic exposure and the induction of human cancers. J Toxicol 2011; 2011431287
[http://dx.doi.org/10.1155/2011/431287] [PMID: 22174709]

[24] Huang C, Ke Q, Costa M, Shi X. Molecular mechanisms of arsenic carcinogenesis. Mol Cell Biochem 2004; 255(1-2): 57-66.
[http://dx.doi.org/10.1023/B:MCBI.0000007261.04684.78] [PMID: 14971646]

[25] Zhou Q, Xi S. A review on arsenic carcinogenesis: Epidemiology, metabolism, genotoxicity and epigenetic changes. Regul Toxicol Pharmacol 2018; 99: 78-88.
[http://dx.doi.org/10.1016/j.yrtph.2018.09.010] [PMID: 30223072]

[26] https://www.cancer.org/cancer/cancer-causes/asbestos.html

[27] Toyokuni S. Mechanisms of asbestos-induced carcinogenesis. Nagoya J Med Sci 2009; 71(1-2): 1-10.
[PMID: 19358470]

[28] Barrett JC, Lamb PW, Wiseman RW. Multiple mechanisms for the carcinogenic effects of asbestos and other mineral fibers. Environ Health Perspect 1989; 81: 81-9.

[http://dx.doi.org/10.1289/ehp.898181] [PMID: 2667990]

[29]   Im S, Youn KW, Shin D, Lee MJ, Choi S-J. Review of carcinogenicity of asbestos and proposal of approval standards of an occupational cancer caused by asbestos in Korea. Ann Occup Environ Med 2015; 27: 34.
[http://dx.doi.org/10.1186/s40557-015-0080-1] [PMID: 26719804]

[30]   Straif K, Baan R, Grosse Y, *et al.* WHO International Agency For Research on Cancer Monograph Working Group. Carcinogenicity of shift-work, painting, and fire-fighting. Lancet Oncol 2007; 8(12): 1065-6.
[http://dx.doi.org/10.1016/S1470-2045(07)70373-X] [PMID: 19271347]

[31]   Falzone L, Marconi A, Loreto C, Franco S, Spandidos DA, Libra M. Occupational exposure to carcinogens: Benzene, pesticides and fibers (Review). Mol Med Rep 2016; 14(5): 4467-74.
[http://dx.doi.org/10.3892/mmr.2016.5791] [PMID: 27748850]

[32]   McMichael AJ. Carcinogenicity of benzene, toluene and xylene: epidemiological and experimental evidence. IARC Sci Publ 1988; (85): 3-18.
[PMID: 3053447]

[33]   Gordon T. Beryllium: Genotoxicity and carcinogenicity Mutat Res Fund Mol M 2003; 533: 99-105.

[34]   Léonard A, Lauwerys R. Mutagenicity, carcinogenicity and teratogenicity of beryllium. Mutat. Res-Gen Tox. En., 1987, 186, 35–42.

[35]   Kuschner M. The carcinogenicity of beryllium. Environ Health Perspect 1981; 40: 101-5.
[http://dx.doi.org/10.1289/ehp.8140101] [PMID: 7023926]

[36]   Waalkes MP. Cadmium carcinogenesis. Mutat Res 2003; 533(1-2): 107-20.
[http://dx.doi.org/10.1016/j.mrfmmm.2003.07.011] [PMID: 14643415]

[37]   Shekhawat K, Chatterjee S, Joshi BJIJAR. Chromium toxicity and its health hazards. Int J Adv Res (Indore) 2015; 3: 167-72.

[38]   Deng Y, Wang M, Tian T, *et al.* The effect of hexavalent chromium on the incidence and mortality of human cancers: A meta-analysis based on published epidemiological cohort studies. Front Oncol 2019; 9: 24.
[http://dx.doi.org/10.3389/fonc.2019.00024] [PMID: 30778374]

[39]   Wang Y, Su H, Gu Y, Song X, Zhao J. Carcinogenicity of chromium and chemoprevention: a brief update. OncoTargets Ther 2017; 10: 4065-79.
[http://dx.doi.org/10.2147/OTT.S139262] [PMID: 28860815]

[40]   Saha SP, Bhalla DK, Whayne TF Jr, Gairola C. Cigarette smoke and adverse health effects: An overview of research trends and future needs. Int J Angiol 2007; 16(3): 77-83.
[http://dx.doi.org/10.1055/s-0031-1278254] [PMID: 22477297]

[41]   Pfeifer GP, Denissenko MF, Olivier M, Tretyakova N, Hecht SS, Hainaut P. Tobacco smoke carcinogens, DNA damage and p53 mutations in smoking-associated cancers. Oncogene 2002; 21(48): 7435-51.
[http://dx.doi.org/10.1038/sj.onc.1205803] [PMID: 12379884]

[42]   Burgio E, Piscitelli P, Migliore L. Ionizing radiation and human health: Reviewing models of exposure and mechanisms of cellular damage. An epigenetic perspective. Int J Environ Res Public Health 2018; 15(9)E1971
[http://dx.doi.org/10.3390/ijerph15091971] [PMID: 30201914]

[43]   Nenoi M, Ed. G., O. Ionizing Radiation Carcinogenesis.Current Topics in Ionizing Radiation Research. InTech 2012.

[44]   Shuryak I, Brenner DJ, Ullrich RL. Radiation-induced carcinogenesis: mechanistically based differences between gamma-rays and neutrons, and interactions with DMBA. PLoS One 2011; 6(12)e28559

[http://dx.doi.org/10.1371/journal.pone.0028559] [PMID: 22194850]

[45]   Dar RA, Shahnawaz M, Qazi PHJTJP. General overview of medicinal plants: A review. J Phytopharmacol 2017; 6: 349-51.

[46]   Cragg GM, Pezzuto JM. Natural products as a vital source for the discovery of cancer chemotherapeutic and chemopreventive agents. Med Princ Pract 2016; 25 (Suppl. 2): 41-59.
[http://dx.doi.org/10.1159/000443404] [PMID: 26679767]

[47]   Iqbal J, Abbasi BA, Mahmood T, *et al.* Plant-derived anticancer agents: A green anticancer approach. Asian Pac J Trop Biomed 2017; 7: 1129-50.
[http://dx.doi.org/10.1016/j.apjtb.2017.10.016]

[48]   Martino E, Casamassima G, Castiglione S, *et al.* Vinca alkaloids and analogues as anti-cancer agents: Looking back, peering ahead. Bioorg Med Chem Lett 2018; 28(17): 2816-26.
[http://dx.doi.org/10.1016/j.bmcl.2018.06.044] [PMID: 30122223]

[49]   Ojima I, Lichtenthal B, Lee S, Wang C, Wang X. Taxane anticancer agents: A patent perspective. Expert Opin Ther Pat 2016; 26(1): 1-20.
[http://dx.doi.org/10.1517/13543776.2016.1111872] [PMID: 26651178]

[50]   Ardalani H, Avan A, Ghayour-Mobarhan M. Podophyllotoxin: A novel potential natural anticancer agent. Avicenna J Phytomed 2017; 7(4): 285-94.
[PMID: 28884079]

[51]   Li F, Jiang T, Li Q, Ling X. Camptothecin (CPT) and its derivatives are known to target topoisomerase I (Top1) as their mechanism of action: did we miss something in CPT analogue molecular targets for treating human disease such as cancer? Am J Cancer Res 2017; 7(12): 2350-94.
[PMID: 29312794]

[52]   Kurek J. Cytotoxic colchicine alkaloids: From plants to drugs.Cytotoxicity. InTech 2018.
[http://dx.doi.org/10.5772/intechopen.72622]

[53]   Zhang T, Chen W, Jiang X, *et al.* Anticancer effects and underlying mechanism of Colchicine on human gastric cancer cell lines *in vitro* and *in vivo.* Biosci Rep 2019; 39(1)BSR20181802
[http://dx.doi.org/10.1042/BSR20181802] [PMID: 30429232]

[54]   Seddigi ZS, Malik MS, Saraswati AP, *et al.* Recent advances in combretastatin based derivatives and prodrugs as antimitotic agents. MedChemComm 2017; 8(8): 1592-603.
[http://dx.doi.org/10.1039/C7MD00227K] [PMID: 30108870]

[55]   Lichota A, Gwozdzinski K. Anticancer activity of natural compounds from plant and marine environment. Int J Mol Sci 2018; 19(11): 3533.
[http://dx.doi.org/10.3390/ijms19113533] [PMID: 30423952]

[56]   Habtemariam S, Lentini G. Plant-derived anticancer agents: Lessons from the pharmacology of geniposide and its aglycone, genipin. Biomedicines 2018; 6(2): 39.
[http://dx.doi.org/10.3390/biomedicines6020039] [PMID: 29587429]

[57]   Philipp S, Sosna J, Plenge J, Kalthoff H, Adam D. Homoharringtonine, a clinically approved anti-leukemia drug, sensitizes tumor cells for TRAIL-induced necroptosis. Cell Commun Signal 2015; 13(1): 25.
[http://dx.doi.org/10.1186/s12964-015-0103-0] [PMID: 25925126]

[58]   Stiborová M, Černá V, Moserová M, Mrízová I, Arlt VM, Frei E. The anticancer drug ellipticine activated with cytochrome P450 mediates DNA damage determining its pharmacological efficiencies: studies with rats, Hepatic Cytochrome P450 Reductase Null (HRN™) mice and pure enzymes. Int J Mol Sci 2014; 16(1): 284-306.
[http://dx.doi.org/10.3390/ijms16010284] [PMID: 25547492]

[59]   Cicenas J, Kalyan K, Sorokinas A, *et al.* Roscovitine in cancer and other diseases. Ann Transl Med 2015; 3(10): 135.
[PMID: 26207228]

[60]   Dey S, Bajaj SO. Promising anticancer drug thapsigargin: A perspective toward the total synthesis. Synth Commun 2018; 48(1): 1-13.
[http://dx.doi.org/10.1080/00397911.2017.1386789]

CHAPTER 8

Basics of Drug Designing Through Small Organic Molecules and Their Toxicological Impact on The Environment

Mohd Azhar Khan[1], Arif Ali[2], Fakhra Jabeen[3], Malik Nasibullah[4], Tahmeena Khan[4], Musheer Ahmad[2,*] and Qazi Inamur Rahman[4]

[1] *K.K.L.Mahavidyala, Kanpur, India*

[2] *Zakir Husain College of Engineering and Technology, Aligarh Muslim University, Aligarh, India*

[3] *Samtah Jazan University, Kingdom of Saudi Arabia.*

[4] *Integral University, Lucknow, India*

Abstract: In the most basic sense, drug design involves designing molecules that are complementary in shape and charge to the biomolecular target with which they interact, and therefore will bind to it. The therapeutic potential of an organic molecule-based chemotherapeutic candidate is influenced by the basic functional groups, where the stereo-arrangement and stereo-selectivity of groups enhance the therapeutic benefits. Stereo-selective organic molecules in different configurations show diverse activity, such as (R) and (S) enantiomers of ibuprofen are effective pain killers but only (S) naproxen has inflammatory activity. Similarly, the transformation of diethyl stilbesterol has potential estrogenic activity and not the cis form. The softness or hardness of drugs depends on the functionality of organic molecules; mostly, the presence of hydroxyl and carboxylic groups improves the softness. This chapter deals with effective drug designing, including the structure-activity relationship and the influence of various functional groups on the activity of a drug compound. The toxicological impact of drugs on the environment has also been explored. In recent times, it has been successfully studied that residue of drugs could enter the ecosystem through the water channel. It directly or indirectly impacts soil, groundwater, and surface water, and creates environmental and health problems.

Keywords: Drug, Environment, Functional group, Organic, Stereochemistry, Therapeutic, Toxicological.

---

* **Corresponding author Musheer Ahmad:** Department of Applied Chemistry, Zakir Husain College of Engineering and Technology, Aligarh Muslim University, Aligarh, India; E-mail: amusheer4@gmail.com

Tahmeena Khan, Abdul Rahman Khan, Saman Raza, Iqbal Azad and Alfred J. Lawrence (Eds.)

INTRODUCTION

Drug Design and Development of New Drugs

Drugs play an essential role in the prevention and treatment of human diseases. Human life is constantly threatened by diseases, such as cancer, and fungal, bacterial, and microbial diseases. Therefore, ideal drugs have great demand for the treatment of diseases. The chemists face the challenge to develop the patent drug design. For meeting these challenges, several multidisciplinary approaches are required for the process of drug development; collectively these approaches form the basis of rational drug designing. Rational drug designing for tailor-made compounds is the inventive process of finding new drugs with a high degree of chemotherapeutic index and specificity of action, based on the knowledge of the biological target [1]. Mostly, the small organic molecules are used to derive drugs, which activate or inhibit the targeted protein function, with a resultant therapeutic benefit to the patient. The selection of organic molecules in drug design depends on their shape and charge concerning the biomolecular target, having its interaction and binding properties [2]. Frequently but not necessarily, computer modelling techniques are used to study the three-dimensional structure of the biomolecular target and are known as structure-based drug design [3]. Mainly, computational methods have been used to enhance the bioavailability of small organic molecules for improving the affinity, selectivity, and stability of protein-based therapeutics. On a large scale, small molecules are an important class of drugs in terms of biopharmaceutical and therapeutic agents. Currently, small molecules or drugs have been showing great efficacy as new cardiovascular agents, antineoplastics, drugs for endocrine diseases, and acting on the central nervous system [4]. Attention should be given to which drug is better for the body and, also what the body does to the drug when it is administered to a patient. The drug must travel a journey in the body before it reaches the target site and performs its work, which is related to pharmacokinetics, and after the drug reaches the target site the mechanism is referred to as pharmacodynamics [5]. During the development of new drugs, drug targets are also identified. These targets include carbohydrates, lipids, proteins, and nucleic acids [4].

The Procedure Followed in Drug Design

The discovery of a new drug is used for the cure or prevention of diseases or the recovery of physical or mental health. Nowadays, several methods are used for rational drug designing [2 - 4]. The principal ones are Computer-Assisted Drug Design (CADD) which is primarily concerned with physicochemical parameters involved in drug activity, quantitative structure-activity relationships (QSAR), and quantum chemistry models (molecular orbital calculations) to determine the most promising drug candidate [3]. Molecular graphics refers to the visualization of molecular objects, and this term is also used as a synonym for molecular modelling. The molecular shape and design of a drug are determined with the help of Single-Crystal X-ray crystallography and spectroscopy techniques [5].

Concept of Lead Compounds and Lead Modifications

A chemical entity may show pharmacological activity to be useful in therapeutics, but some optimal properties are required in the drug structure for better interaction with the target [6]. The chemical structure is used for chemical modifications to improve the potency, selectivity, and pharmacokinetic parameters of drugs [7]. Furthermore, the newly invented organic molecule may have poor drug activity and may require chemical modification to become a pharmacologically active drug, which is confirmed on basis of biological or clinical tests [8]. Natural products are a rich source of potential lead compounds. Plants, snakes, lizards, frogs, fungi, corals, and fish have all yielded potent lead compounds which have either resulted in clinically useful drugs or have the potential to do so. Recently, drug discovery has significantly increased due to the availability of 3D X-ray or NMR structures of biomolecules and docking tools, and the development of computer-aided methodologies.

Concept of a Prodrug, Double Prodrug, and Soft Drug

a. Prodrugs

After administration, prodrugs are used as pharmacologically active drugs. A prodrug is metabolized within the body and may be used to improve how a medicine is absorbed, distributed, metabolized, and excreted (ADME) [9]. Prodrugs are often designed to improve bioavailability when a drug itself is poorly absorbed from the gastrointestinal tract [10], especially in chemotherapy, it can reduce the adverse or unintended side effects of the drug [11].

Drug latentation consists of chemically transforming an active drug (parent drug) to an inactive form, which is converted to parent drug within the body before exhibiting its pharmacological action. The latent form of the drug is known as prodrug [12]. Prodrugs are prepared to minimize the unpleasant odour and taste, pain at the site of injection, and gastrointestinal irritation. For example, to mask the bitter taste of the chloramphenicol and lincomycin, they were converted to the form of palmitates, which are tasteless and release the active antibiotics *in vivo* [13].

A few drugs do not exert any physiological action in the body and require conversion in the body to give one or more active metabolites, such drugs are called prodrugs. A substance that takes part in cellular metabolic reactions is known as a metabolite, while a chemical agent which blocks the metabolism due to its close structural similarity to the metabolite is known as antimetabolite [11].

For the better bioavailability of a drug, the prodrug should have no side effect, should be less toxic and more stable, and should possess desirable pharmacokinetic properties. For example, the prodrug of active prednisolone is prednisone and the prodrug of active ampicillin is bacampicillin [14a].

Another example of a prodrug is an ester that produces active alcohol *in vivo*. The rate of the release of the alcohol can be controlled by the electronic and steric nature of the carrier.

i. *Electronic nature*

Electronic nature controlling the release of alcohol is presented in equation **1**:

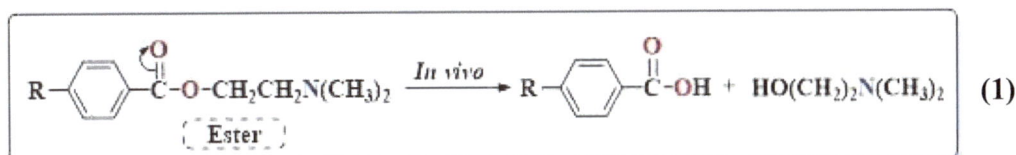

$$R-\langle\text{aromatic}\rangle-\overset{O}{\overset{||}{C}}-O-CH_2CH_2N(CH_3)_2 \xrightarrow{\textit{In vivo}} R-\langle\text{aromatic}\rangle-\overset{O}{\overset{||}{C}}-OH + HO(CH_2)_2N(CH_3)_2 \quad (1)$$

$\boxed{\text{Ester}}$

ii. *Steric Factors*

The release of alcohol is controlled by bulky groups, as presented in equation **2**:

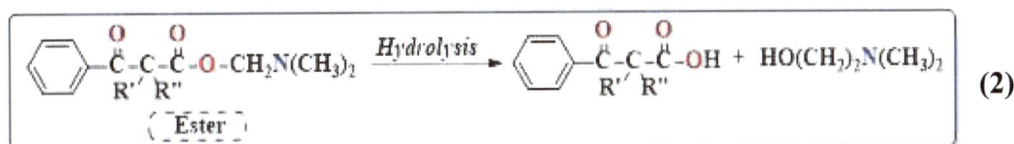

$$\langle\text{aromatic}\rangle-\overset{O}{\overset{||}{C}}-\underset{R''}{\overset{R'}{C}}-\overset{O}{\overset{||}{C}}-O-CH_2N(CH_3)_2 \xrightarrow{\textit{Hydrolysis}} \langle\text{aromatic}\rangle-\overset{O}{\overset{||}{C}}-\underset{R''}{\overset{R'}{C}}-\overset{O}{\overset{||}{C}}-OH + HO(CH_2)_2N(CH_3)_2 \quad (2)$$

$\boxed{\text{Ester}}$

If R′ and R′′ both are H, the hydrolysis rate (μg/ml of serum/hr) is found to be 500, if R′ and R′′ both are CH_3 then hydrolysis rate is found to be only 35 [14 b].

Esters (prodrugs) also produce active mercaptans. The presence of sulphur in mercaptans enhances the chemical reactivity because sulphur is more reactive than oxygen (Eq. **3**). Due to this factor, sulphur containing active compound, ethanethiol, has good antileprotic and antitubercular properties; it has an unpleasant odour, appearing in the breath.

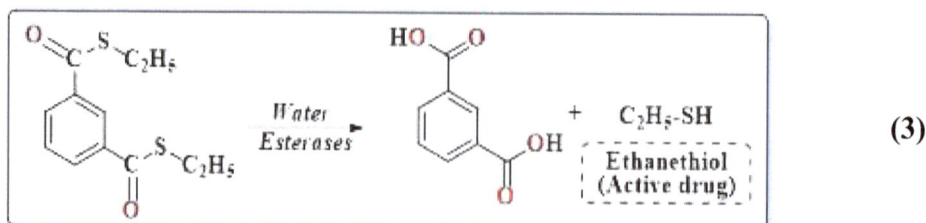

$$(3)$$

Amino acid-containing Levodopa, which is converted into dopamine in the brain by decarboxylase enzyme, is a prodrug while dopamine is its active form (Eq. **4**).

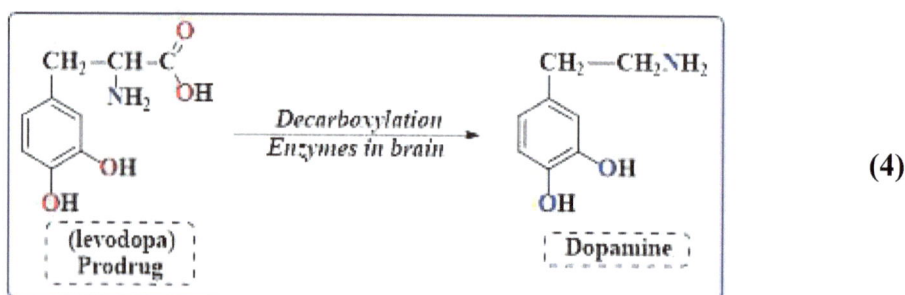

$$(4)$$

b. Double Prodrugs

A double prodrug has two ester groups in a molecule and is biologically inactive; when this is transformed *in vivo* enzymatically and/or chemically then it converts into an active species [12]. In the case of the double prodrug concept, the cleavage of the first ester group is enzymatic (Eq. **5**), while hydrolysis of the second ester group is non-enzymatic *i.e.*, by chemical means.

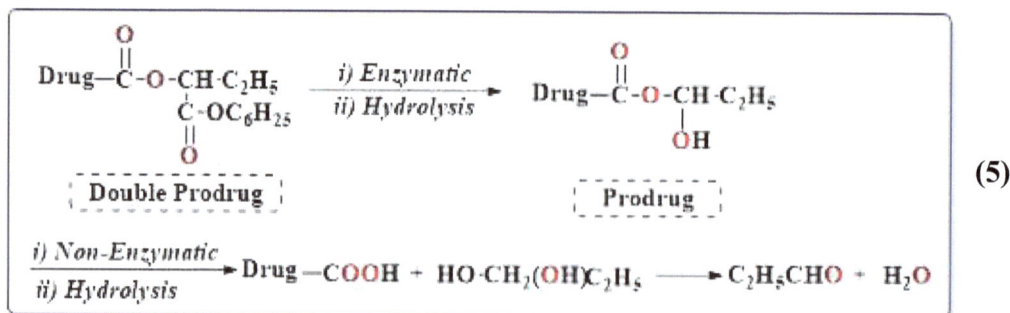

$$(5)$$

c. Soft Drugs

A soft drug can be:

- A drug with low physical or psychological dependence as compared to hard drugs.
- A metabolically unstable drug that is rapidly cleared (synonym: antidrug; antonym: prodrug).

The term 'soft drug' has been derived based on hard and soft acids and bases. A soft drug is easily metabolized, whereas a hard drug means a compound which is hard to metabolize or is non-metabolisable [14]. The molecule of a soft drug is physiologically active. The main purpose of preparing soft drugs is to increase the therapeutic indices, not to enhance the potency (Fig. **1**).

$$TI = \frac{TD_{50}}{ED_{50}}$$

Fig. (1). Therapeutic indices, where TI= Therapeutic Index, TD_{50} = median toxic dose & ED_{50}= median effective dose.

A prodrug is inactive and may be activated *in vivo* to the active drug, while the soft drug is the active species. It is possible to prepare a "pro-soft drug" but a "soft prodrug" does not exist [15].

Soft drugs are divided by Nicolas Bodor into five different groups: soft analogues, activated soft compounds, natural soft drugs, soft drugs based on the active metabolite approach and soft drugs based on the inactive metabolite approach.

1. Soft Analogues

Soft analogues are 'close structure' analogues of known active drugs or bioactive compounds which have a specific metabolically (preferably) sensitive spot built into their structure, which provides their one-step, controllable detoxication. These sensitive structural parts are not oxidizable alkyl chains or functional groups subjected to conjugation. The designed detoxification takes place as soon as the desired activity is achieved, not allowing other types of metabolic transformations to take place. The simplest example of a soft analogue is the isosteric analogue (II) of Cetylpyridinium chloride (I), which is a 'hard' quaternary antimicrobial agent (Fig. **2(a),(b)**). Gallic acid is a chemical compound that is naturally occurring and has anticarcinogenic, anti-microbial, anti-mutagenic and antioxidant effects and predicts the solubility and the percentage of absorption for the designed drug candidates multi-target antibiotic effect with sulfamethoxazole (Fig. **3**).

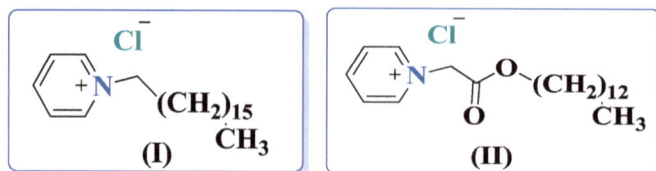

Fig. (2). (a) Isosteric analogue of cetylpyridinium chloride, (b) cetylpyridinium chloride.

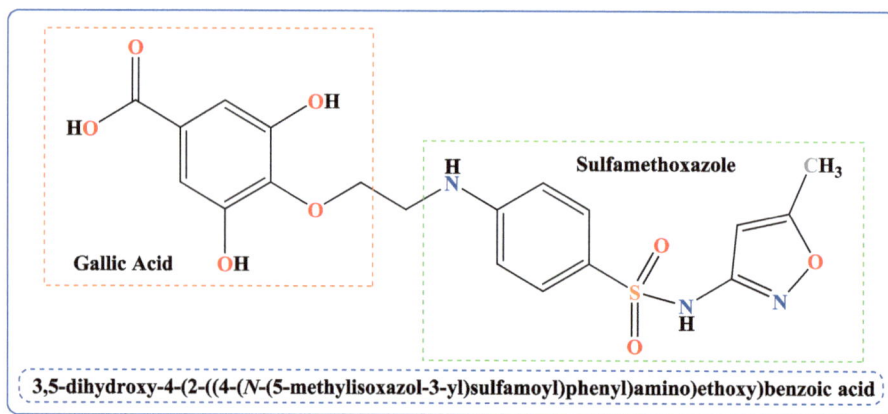

Fig. (3). Multi-target Antibiotic Phyto drug conjugates.

2. Activated Soft Compounds

They are designed by introducing a pharmacophoric group in the structure of a nontoxic, inactive compound to activate it to exhibit a certain pharmacological activity [17]. *In vivo,* the activated form loses the activating group and reverts to the original non-toxic compound, and/ or undergoes further metabolism to non-toxic moieties while performing its role.

3. Natural Soft Drug

Several endogenous substances can be considered as natural soft drugs since the body possesses efficient and fast metabolic pathways for their disposal, without going through highly reactive intermediates. In other words, their metabolism is predictable [18]. When employed at concentrations close to their normal levels, they do not produce unexpected toxicity. Example: (a) neurotransmitters: γ-aminobutyric acid, dopamine, epinephrine, and norepinephrine; and (b) steroid hormones: estradiol, hydrocortisone, progesterone, and testosterone. Their therapeutic index can be improved by using chemical delivery systems to achieve sustained, local, or site-specific delivery. L-DOPA drug is metabolically converted into dopamine which acts as an active therapeutic agent (Figs. **4-5**).

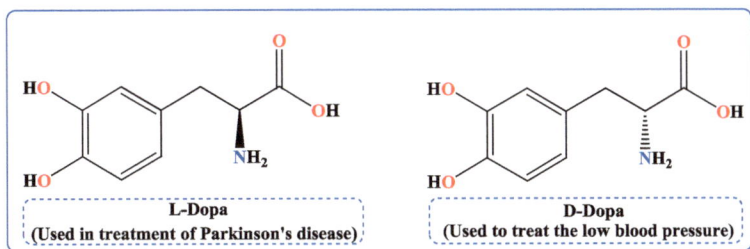

Fig. (4). Structure of L and D Dopa.

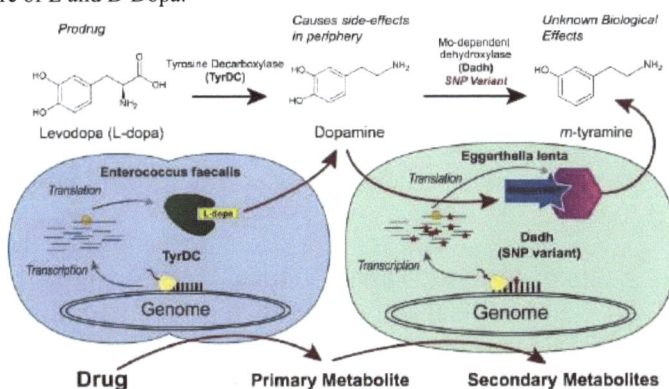

Fig. (5). Mechanism of DOPA drug.

4. Soft Drug Based on the Active Metabolite Approach

Some drugs undergo stepwise biotransformation, giving intermediates and structural analogues which have an activity similar to that of the original drug molecule [17]. According to Bodor, 'It is preferable to use as the drug of choice, an active species which undergoes a one-step, singular, predictable metabolic deactivation, whenever oxidative metabolic transformation of a drug takes place, going through possibly toxic, highly reactive intermediates or pharmacologically active species, if activity and pharmacokinetic considerations permit it, the drug of choice should be the active metabolite which is in the highest oxidized state'.

5. Soft Drug Based on the Inactive Metabolite Approach

The inactive metabolite approach is one of the most promising and most versatile methods for designing safe soft drugs [16]. Preparation of soft drugs by this approach involves three stages, (a) activation stage- chemical modification of a known inactive metabolite of a drug (by isosterism); this metabolite is used as the lead compound; (b) predictable metabolism- design of the structure of the new soft analogue in such a way that its metabolism will yield the starting inactive metabolite in one step and without going through toxic intermediates; and (c) controllable metabolism- control of transport and binding properties as well as the rate of metabolism and pharmacokinetics by molecular modifications in the first, activation stage.

Structure-Activity Relationship (SAR)

The structure-activity relationship (SAR) is the relationship between the chemical or 3D structure of a molecule and its biological activity [18]. The analysis of SAR enables the determination of the chemical groups responsible for evoking a target biological effect in the organism [19]. This allows modification of the effect or the potency of a bioactive compound (typically a drug) by changing its chemical structure. Medicinal chemists use chemical synthesis to insert new chemical groups into the medicinal compounds and test the modifications for their biological effects [20].

This method was refined to build mathematical relationships between the chemical structure and the biological activity, known as the quantitative structure-activity relationship (QSAR).

It was found that the physiological activity of a compound is associated with a specific structural group. The part of the compound which is responsible for the actual physiological activity is known as the pharmacophore group. This pharmacophore group, if modified by the simple and common processes, gives a more physiologically active compound with low toxicity [19]. The important groups along with their physiological effects are explained below:

1. Effect of the Hydroxyl Group

Introduction of the hydroxyl group into aliphatic compounds generally decreases their biological and physiological activity; for example, propanol is much more active than glycerol. The physiological action of caffeine is lost in hydroxy-caffeine. In aromatic compounds, the physiological effects of the hydroxyl group are reversed as compared to that of aliphatic compounds [20]. The addition of hydroxyl groups generally increases the physiological action of aromatic compounds. For example, salicylic acid shows antiseptic and antirheumatic properties as compared to the inert potent compound, benzoic acid. Phenol is an antiseptic compound and shows more toxicity than benzene. Polyphenols are more toxic than phenol.

2. Effect of Aldehydes and Ketones

Aldehydes are more reactive than ketones. Formaldehyde is an antiseptic compound and exerts a hardening effect on tissues. The introduction of hydroxyl groups in an aldehyde compound causes a reduction in physiological action and sometimes, the compound becomes medicinally inert, for example, glucose.

3. Effect of Acidic Groups

The introduction of an acidic group in a compound either decreases or removes the biological action of the parent compound. For example, phenol is poisonous but benzene sulphonic acid is harmless. Aniline is toxic while meta-aminobenzoic acid causes no harm. Amines are toxic compounds, whereas amino acids are used as foodstuffs. Morphine shows strong biological activity but morphine sulphuric acid is completely inactive.

4. Effect of Alkyl Groups

It has been observed that the biological activity of the alkylated compound is less as compared to the non-alkylated compound. For example, the introduction of an alkyl group in ammonia reduces its convulsive properties. There are many exceptions to the above rule. In several cases, the introduction of an alkyl group increases the biological activity of the compound. For example, antipyrine (Fig. **6, 7**) is a strong antipyretic, while a reduction in the number of methyl groups leads to its inactivity. Cocaine is a strong anaesthetic drug whereas its analogous acid is inert (Fig. **8, 9**).

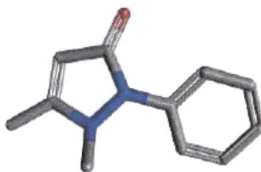

Fig. (6). Ball and stick model of Antipyrine ($C_{11}H_{12}N_2O$).

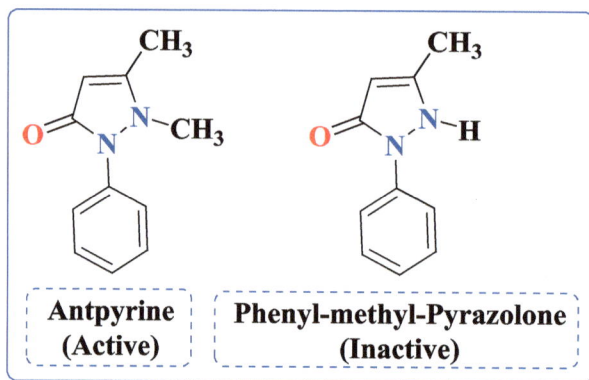

Fig. (7). Structures of antipyrine and phenyl-methyl-pyrazole.

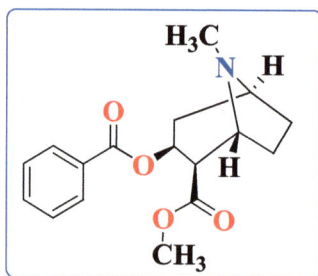

Fig. (8). Structure of cocaine ($C_{17}H_{21}NO_4$).

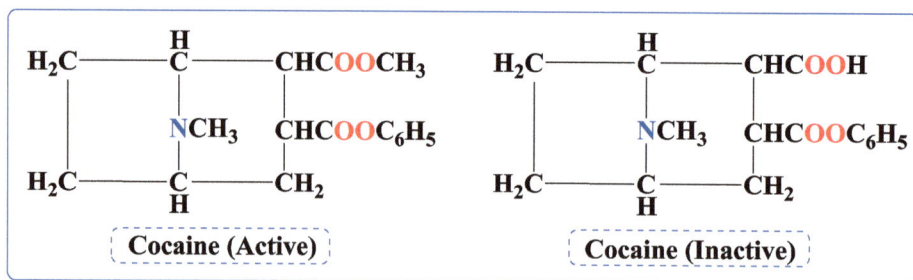

Fig. (9). Active and inactive cocaine.

Analogues of lead compounds are synthesized and tested to determine how structural variations affect pharmacological activity. One of the most useful bits of information that can be obtained from SAR studies is the type of atoms and functional groups that are important in the binding of a drug to its target binding site. Usually, such bindings are intermolecular bonding like ionic bonds, hydrogen bonds, van der Waals interactions, and dipole-dipole interactions, as in some cases drugs covalently bind with their targets.

History and Development of QSAR (Quantitative Structure-activity Relationship)

It has been nearly 40 years since the quantitative structure-activity relationship (QSAR) paradigm first found its way into the practice of agro-chemistry, pharmaceutical chemistry, toxicology, and eventually, most facets of chemistry [21, 22]. The strategy works on the basic principle that activity is a function of structure as described by electronic attributes, hydrophobicity, and steric properties, as well as the rapid and extensive development in methodologies and computational techniques. The overall goals of QSAR retain their original essence and remain focused on the predictive ability of the approach and its receptiveness to mechanistic interpretation. Rigorous analysis and fine-tuning of independent variables have led to an expansion in the development of molecular and atom-based descriptors, as well as descriptors derived from quantum chemical calculations and spectroscopy [23].

Scientists involved in rational drug design should possess a considerable amount of imagination as well as a stochastic mind, that is, they should be apt to divine the truth by conjecture [23]. Furthermore, they must resort to up-to-date progress of several sciences contributing to this area, such as chemistry, biochemistry, biology, psychology, microbiology, parasitology, immunology, and pharmacology. In short, rational drug designing consists of utilizing the new available scientific knowledge. Generally, QSAR is dependent on the polarity and/or ionization of the drug, which can be altered or changed by its substituents; these changes alter quantitative structure-activity relationships (QSAR) [23].

a. The site and mechanism of drug action at molecular and electronic levels.
b. Qualitative and quantitative structure-activity relationship.
c. Drug receptors and their tridimensional topography.
d. Mode of drug-receptor interactions.

Traditional approaches such as trial-and-error synthesis of compounds and random screening for activity are time-consuming and expensive. It is required that the synthesis of such compounds be carefully designed and their biological activity should be determined on suitable test systems [24].

From QSAR, it may be possible to elucidate the influence of various physicochemical properties on drug potency.

Quantitative Structure-Activity Relationship (QSAR) methods

1. Discriminant analysis
2. Free-Wilson or the additive model
3. Hansch approach
4. Molecular modelling
5. Molecular orbital method
6. Principal component analysis (PCA)

1. Free Wilson model

This is based on the assumption that the biological activity of a molecule is the sum of the activity contributions of definite structures *i.e.*, of the parent fragment (unsubstituted) and the substituents (Eq. **6**).

$$log\ (1/C) = \sum Gij + \mu \dots ij \tag{6}$$

μ is the overall average of the biological activity, and Gij is the activity contribution of the substituent at position j.

2. Hansch approach

Hansch equation is the QSAR equation (Eq. **7**) that relates physicochemical properties to activity including a variety of parameters, the most common being log P, π, σ, F, R, MR and Es.

$$log\ (1/C) = -K_1(logP)^2 + K_2 logP + K_3\pi + K_4\sigma + K_5 ES + K_6 \tag{7}$$

Where K_1-K_6 are the constants and are calculated computationally.

Here, the variance in biological activity (1/C) is explained by the linear free energy-related substituent constants which describe the variance in lipophilic/hydrophilic, electronic, steric, or other properties of the parent molecule induced by the substituent.

The change in lipophilicity can be described by the partition coefficient, log P, or the substituent constant defined as π = log Px-log PH, where 'x' refers to the substituted derivative and 'H' to the parent compound. The change in electronic properties can be expressed by Hammet constants, pKa values, and spectroscopic values like NMR, IR, and UV.

The steric influence of the substituents can be expressed by the Taft steric constant (Es), the van der Waals volume, or the molar volume (MV) and molar refractivity.

Relationship Between free-Wilson and Hansch Analysis

Kubinyi combined Hansch and Free Wilson models as a 'mixed approach'. This is based on the theoretical and numerical equivalence of Hansch's linear multiple regression model and modified free Wilson model and represents combinations of both these models (Eq. **8**).

$$log\ (1/C) = \sum ai + \sum kj\phi j + k \ ...$$
(8)

Factors Affecting Bio-activity

Spatial Considerations

The branch of science which deals with the three-dimensional (3D) arrangement of atoms in space is defined as stereochemistry. Stereoisomerism is divided into the following types.

1. Optical Isomerism

For a compound to show optical activity, 'chirality' is a must; an object lacking an element of symmetry is called asymmetric or chiral [25]. While discussing optical isomerism we must distinguish between dextrorotatory or (+) and laevorotatory or (-).

R, S System for Asymmetric Molecules

In R, S system (Absolute configuration or Cohn-Ingold-Prelog rule CIP), we arrange four groups attached to the asymmetric carbon in decreasing order of priority by applying sequence rules. Racemic ibuprofen is used as an anti-inflammatory and analgesic agent, whereas S-ibuprofen is an active pain killer and capable of inhibiting cyclooxygenase (COX), and R-ibuprofen is not a COX inhibitor (Fig. **10**).

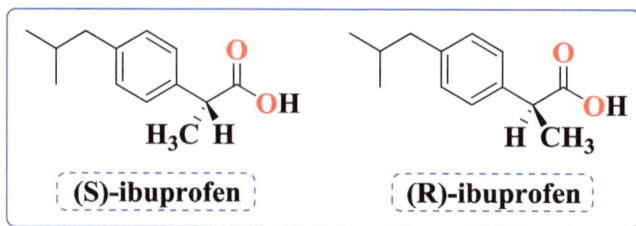

Fig. (10). Structure of S and R ibuprofen.

Optical Isomerism and Biological Activity

(-) Ephedrine shows 36 times more activity than (+) pseudoephedrine (Fig. **11**). The (+) enantiomer of limonene has an orange-like odour, whereas the (-) enantiomer of limonene has a lemon-like odour. Both naproxen and ibuprofen are chiral, but while both enantiomers of ibuprofen are effective pain killers and the drug is sold as a racemic mixture, only (S) enantiomer of naproxen has anti-inflammatory activity. Similarly, propoxyphene isomers have different properties (Fig. **12**).

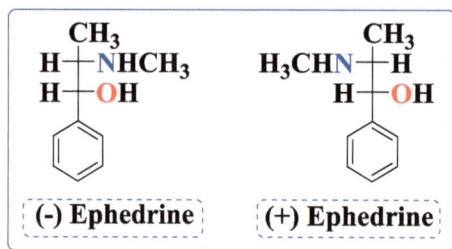

Fig. (11). Structure of Ephedrine.

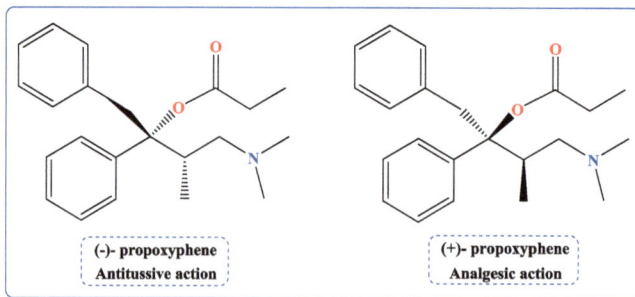

Fig. (12). Isomers of Ephedrine and propoxyphene drug.

2. Geometrical Isomerism

Cis/trans isomerism, also known as geometric isomerism or configurational isomerism, is a term used in organic chemistry to refer to the stereoisomerism related to the relative orientation of functional groups within a molecule [26]. It is not to be confused with E/Z isomerism (Fig. **13**), which is an absolute stereochemical description. In general, such isomers contain double bonds that cannot rotate, or they may contain ring structures, where the rotation of bonds is restricted or eliminated. Cis and trans isomers occur both in organic molecules and in inorganic coordination complexes [27]. Cis and trans descriptors are not used for cases of conformational isomerism where the two geometric forms easily interconvert, such as most open-chain single-bonded structures; instead, the terms 'syn' and 'anti' would be used.

The trans isomers are more stable than cis-isomers because in cis-isomers the bulky groups are on the same side of the double bond (Fig. **13, 14**). Trans geometrical isomer has no dipole, so it is easily separated by fraction distillation, gas chromatography, and other spectroscopic techniques [26]. Also, cis acid is found to be stronger than the corresponding trans acid (Fig. **13, 14**).

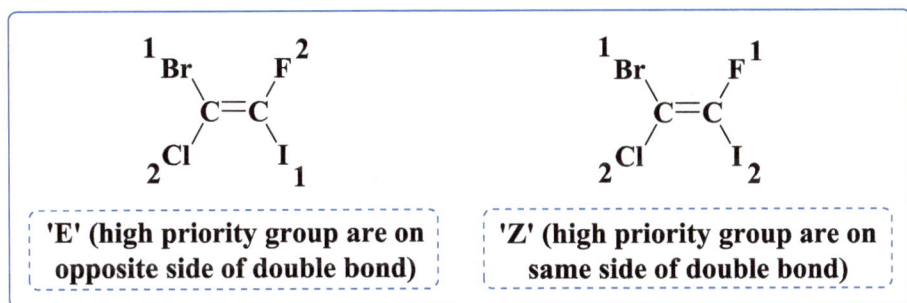

Fig. (13). E and Z isomer.

Priority Order Based on Atomic Number

$I > Br > Cl > SO_3H > SH > F > OR > OH > NO_2 > NR_2 > NHCOR > NH_2 > CCl_3 > COCl > COOR > COOH > CONH_2 > CHO > CH_2OH > CN > CR_3 > C_6H_5 > CH_2R > CH_3 > D > H.$

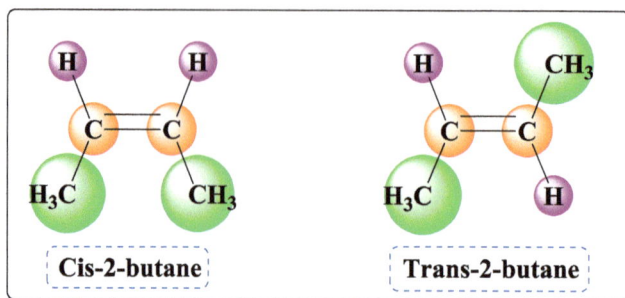

Fig. (14). Cis, trans isomers of 2-butene.

Geometrical Isomerism and Pharmacological Activity

Trans form of diethyl stilbesterol has fourteen times the estrogenic activity than the cis isomer (Fig. **15**).

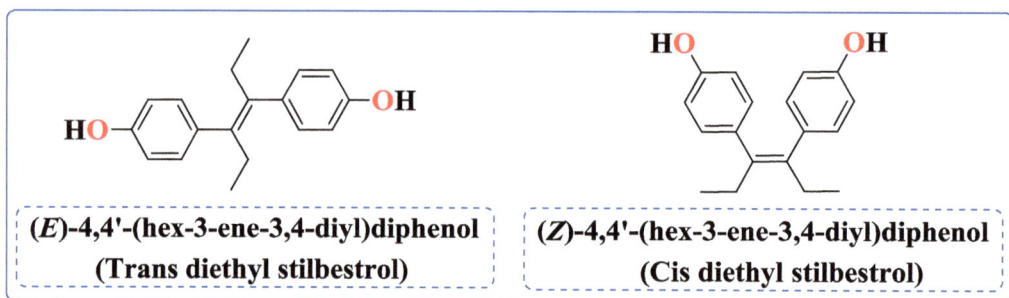

Fig. (15). Geometrical isomers of diethyl stilbestrol.

Isosterism

A group of atoms, molecules, or radicals that have similar physiochemical properties due to similar electronic structure, are known as 'isosteres' and this phenomenon is called 'isosterism'. For example, the chemical properties of N_2 and CO are similar and the chemical properties of N_3^-(azide) and $-N=C=O$ (isocyanide) are similar (Table **1**).

Table 1. Isosteres.

Isosteres	No. of Electrons
CO	6+8=14
N_2	7+7=14
N=C=O	7+6+8=21
N_2O	7+7+8=22

(Table 1) cont.....

Isosteres	No. of Electrons
N_3^-	$3x7+1=22$
CO_2	$6+8+8=22$

Bio-isosterism

In medicinal chemistry, bioisosteres are chemical substituents or groups with similar physical or chemical properties which produce broadly similar biological properties [23]. In drug design, the purpose of exchanging one bioisostere for another is to enhance the desired biological or physical properties of a compound without making significant changes in chemical structure [28]. The main use of this term and its techniques are related to pharmaceutical sciences. Bioisosterism is used to reduce toxicity, change bioavailability, or modify the activity of the lead compound, and may alter the metabolism of the lead. Bioisosteres are the isosteric compounds that have the same type of biological activity. Diethyl stilbestrol and estradiol are classic examples of cyclic and non-cyclic bioisosteres (Fig. **16, 17**). Diethyl stilbestrol has the same potency as natural estradiol. Isosteric replacement of cyclic *versus* non-cyclic analogues produces bio-isosteres (Fig. **18**). Isosteric replacement of groups also gives bioisosteres (Fig. **19**).

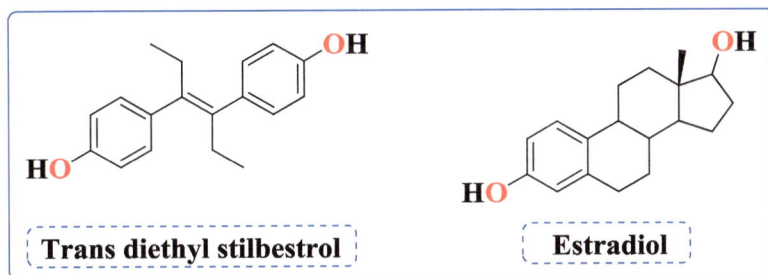

Fig. (16). Stilberstrol and estradiol.

Fig. (17). Spacefilling model of estradiol showing Isoesteric replacement of Si for C to make bio-isostere.

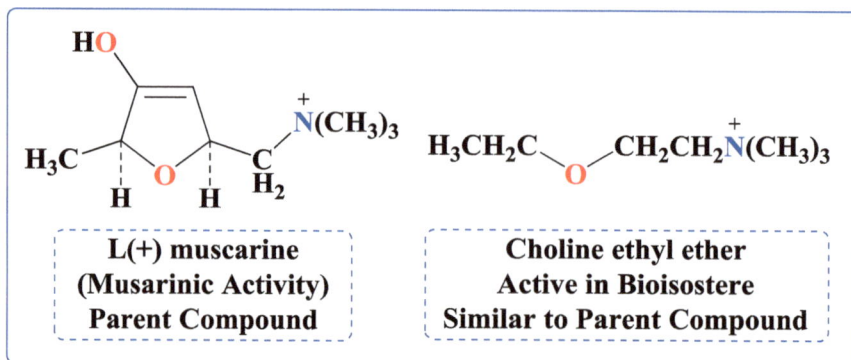

Fig. (18). Isosteric replacement in L(+) muscarine.

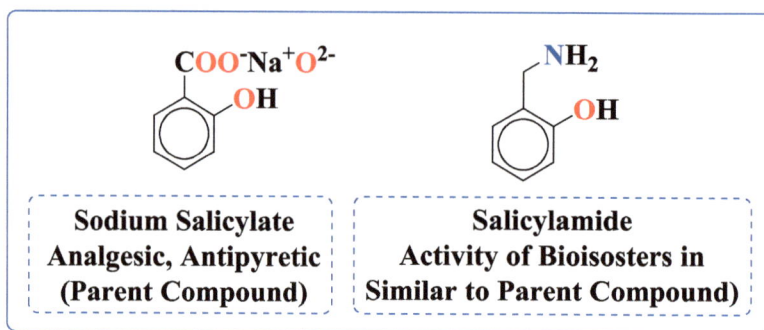

Fig. (19). Isosteric replacement in sodium salicylate.

Redox Potentials

Reduction potential (also known as redox potential, oxidation/reduction potential, ORP, pE, ε, or E_h) is a measure of the tendency of a chemical species to acquire electrons and thereby be reduced. Reduction potential is measured in volts (V), or millivolts (mV). Each species has its intrinsic reduction potential; the more positive the potential, the greater the species' affinity for electrons and tendency to be reduced.

Based on the mode of pharmacological action, drugs may be structurally non-specific and structurally specific. Structurally non-specific drugs are those in which pharmacological action is not directly related to the chemical structure but to other properties like adsorption, solubility, pKa, and oxidation-reduction potential which influences permeability, depolarization of the membrane protein coagulation, and complex formation.

Table 2. Isonarcotic concentrations for mice and thermodynamic activity of drugs.

Drugs	P	Ps	N	Pt	A
Acetylene	1.8	51700	65	494	0.01
Chloroform	265	324	0.5	4	0.01
Methyl ether	11.5	6100	12	91	0.02
Nitrous oxide	1.4	59300	100	760	0.01

P = Partition coefficient (oil: vapour) Ps =Saturated vapour pressure (mm) at 37 ^{0}C N = Norcotic concentration (% by volume) Pt= Partial pressure (mm) at narcotic concentration (760×C/100), a = thermodynamic activity; Pt/Ps

Structurally specific drugs are those whose biological action results essentially from their chemical structure, which should adapt themselves to the three-dimensional structure of receptors in the organism by forming a complex with them.

Hence, in these drugs chemical reactivity, shape, size, the stereochemical arrangement of the molecule, and distribution of functional groups, as well as resonance, inductive effects, electronic distribution, and possible binding with receptors, besides other factors, play a decisive role.

Theories of Drug Activity

To determine molecular mechanics, many mathematical approaches are used for the computational study of structure, dipole moment, energy, and other physical and biochemical properties. In general, quantum mechanics/ molecular mechanics (QM/MM) are utilized.

One mole of a drug contains 6.023×10^{23} molecules. If the molecular weight of the drug is 200 mg, then 1 mg (often the effective dose) will contain $6 \times 10^{23} = 3 \times 10^{18}$ molecules *i.e.* 200×10^3. The human organs are made up of about 3×10^{13} cells. One erythrocyte is found to contain approximately 1×10^{10} molecules. Therefore, the human organs contain approximately, $3 \times 10^{18} \times 10^{10} = 3 \times 10^{28}$ molecules. Consequently, each drug molecule would act on 100,000 molecules of the human body.

This is impossible because the drug molecule cannot be divided into such a large number of parts [31].

Mainly, the drug activity depends on physicochemical kinetics and thermodynamics. The effective molarity of a drug shows the efficiency of an intermolecular process is given by equation **9** [14],

$$logEM_{cal} = 0.809logEM_{exp} + 4.75 \qquad (9)$$

Where, EM_{cal} = calculated effective molarity.

EM_{exp} = Experimental effective molarity.

1. Receptor Theory

Receptor theory is the application of receptor models to explain drug behaviour. Pharmacological receptor models predicted accurate knowledge of receptors for many years. John Newport Langley and Paul Ehrlich introduced the concept of a receptor that would mediate drug action, at the beginning of the 20th century. A. J. Clark was the first to quantify drug-induced biological responses (using an equation described first by A. V. Hill in 1909 and then in 1910) and proposed a model to explain drug-mediated receptor activation [31, 32]. So far, nearly all of the quantitative theoretical modelling of receptor function has centred on ligand-gated ion channels and GPCRs. Postulates of receptor theory are-

• Receptors must possess structural and steric specificity.
• Receptors are saturable and finite (limited number of binding sites).
• Receptors must possess a high affinity for endogenous ligands at physiological concentrations.
• Once the endogenous ligand binds to the receptor, some early recognizable chemical event must occur.

2. Occupancy Theory

According to the occupancy theory, drug-receptor interactions, comply with the law of mass action, as shown by equation **10.**

Where, R = receptor and D = a molecule of the drug.

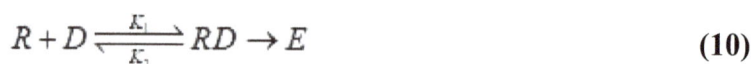

$$R + D \underset{K_2}{\overset{K_1}{\rightleftharpoons}} RD \rightarrow E \qquad (10)$$

RD = Drug-receptor complex

E = Pharmacological effect

K_1, K_2 = Rate constants of absorption and desorption respectively.

3. Affinity and Intrinsic Activity

For a chemical compound to manifest biological activity, it should not only have the affinity for the receptor, owing to complementary structural characteristics, but also intrinsic activity or efficiency. The latter property would be a measure of the ability of the drug-receptor complex to produce the biological effect [33]. According to the Ariens-Stephensen theory, agonists, as well as antagonists, have a strong affinity for the receptor, and this enables them to form the drug-receptor complex. However, only the agonist can give origin to the stimulus. Agonists are usually small molecules containing polar groups. In epinephrine, for instance, such polar groups are the amino, the ß-hydroxyl, and the hydroxyls of the catechol nucleus. Mainly, aromatic rings that establish strong interaction with receptors bind to receptor accessory areas and thereby block the action of agonists. Rang and Ritter have shown that agonists may change the molecular structure of the receptor and this change increases the affinity of the receptor towards some antagonists.

4. Charniere Theory

To explain why an agonist, although not being able to remove an antagonist from the receptor site, competes with it according to the law of mass action, R. Silva proposed the Charniere theory. According to this theory, there are two sites in a pharmacological receptor: (a) a specific or critical site, which interacts with the pharmacophoric groups of the agonist, and (b) a non-specific or non-critical site, which complexes with the non-polar groups of the antagonist. Both agonist and antagonist fix to the specific site through weak reversible bonds, but the antagonist also binds strongly through hydrophobic and van der Waals interactions, as well as charge transfer, to the non-agonist specific site [34]. Competition occurs at the specific site of the receptor. Since antagonist complexes steadily with the non-specific site of the receptor, even an excess of agonist is unable to dislodge it from there.

5. Rate Theory

In the case of agonists, the rate of association and dissociation are fast (the dissociation is faster than association) [35]. When we consider antagonists, however, the rate of association is faster, but the rate of dissociation is slow.

6. Induced-fit-theory

It is applied to drug-receptor interaction. According to this theory, the biological effect produced by drugs results from the activation or deactivation of enzymes or even non-catalytic proteins, through a change in the tertiary structure of enzymes or proteins [36]. Drugs that have a flexible structure can also undergo conformational change as they approach the site of action or the receptor site; this triggers the stimulus, which leads to biological effect.

Concepts of Drug Receptors

Some drugs exhibit biological activity in minute concentrations. For this reason, they are described as structurally specific. The effect produced by them is attributed to the interaction with a specific cellular component, known as a receptor [37]. As a result of this interaction, the drug forms a complex with the receptor. Receptors are an integral part of certain molecules of living organisms. In most cases, they are the segments of proteins (Table **3**).

Table 3. Receptors, acceptors, or site of action of some drugs.

Drugs	Receptor or Site of Action
Aspirin	Prostaglandins synthase-I
Cardiac glycosides	Na^+/K^+-transporting ATPase
Chloroquine	DNA
Local anaesthetics	Cell membrane
Salicylates	Prostaglandins synthase
Streptomycin	Ribosomes

Drug-Receptor Interaction

Specific drugs attach themselves to their receptors through the same forces that are involved in interactions between simple molecules. Table **4** not only lists the forces responsible for the formation of the drug-receptor complex but also presents some typical examples of their effects. The higher the variation of energy, the higher the proportion of atoms in bonded form.

Physico-Chemical Parameters

Several mathematical equations have been devised that co-relate chemical structure with pharmacological activity. In these equations, certain parameters are considered which represent the physicochemical properties of drugs [38]. More than 80 parameters have been used and can be classified into four categories: solubility, empirical, electronic, non-empirical electronic, and steric parameters.

Table 4. Free energies involved with different bond types.

Bond Type	Free Energy(kJ/mol)	Example
Covalent bond	-(170-460)	$RO - COR$
Ionic bond	-(20)	$R_4N^+I^-$
Hydrogen bond	-(4-30)	$-OH\ldots\ldots O = C$
Charge transfer	-4(4-30)	$OH----\ \overset{C}{\underset{C}{\mid\mid}}$
van der Waals interaction	- (2-4)	$C------C$

Solubility Parameters or Lipophilic Parameters

Also called lipophilic or hydrophobic parameters, they are related to the possibility of attraction and interaction between hydrophobic regions of drugs and receptors. The main parameters of this class are partition coefficient, surface tension, electronic charge density, and surface activity.

Partition Coefficient Parameter

The biological activity of several groups of compounds can be correlated with their partition coefficients in polar and non-polar solvents. Overtone and Meyer were the pioneers in these studies. They resorted to partition coefficients first, to explain the activity of certain narcotics and later, of general anaesthetics.

Surface Activity Parameters

Lipophobic, hydrophilic, or polar groups in order of decreasing efficiency are -OSO_2ONa, -$COONa$, -SO_2Na, -OSO_2H, and -SO_2H. The other less efficient groups are $-OH$, -SH, -O-, $C=O$. $-CHO$, -NO_2, NH_2-NHR, -NR_2, -CN, -CNS, -$COOH$, -$COOR$, -OPO_3H_2 Lipophilic, hydrophobic, or non-polar groups increase the lipid solubility of the compounds of which they are part. Examples of these groups are aliphatic hydrocarbon chains, aryl alkyl groups, and polycyclic hydrocarbon groups [39]. Surfactant agents present two distinct regions of lipophilic and hydrophilic character and are used mainly as detergent, wetting, foaming, and emulsifying agents. Their greater or lesser hydrophilicity and lipophilicity depend on the degree of polarity of the group present. They can be divided into non-ionic, cationic, anionic, and amphoteric categories. Non-ionic surfactants are widely used in pharmaceutical products for the oral use of solubilizing agents of water-insoluble or slightly soluble drugs.

Empirical Electronic Parameters

pKa

The partial lipid nature of cellular membranes, such as the ones that enwrap the stomach, small intestine, mucosa, and nervous tissue, facilitates the passage of drugs with high lipid solubility across them. This lipid solubility is affected by the pH of the environmental medium and by the degree of dissociation pK_a.

Usually, drugs are weak acids or weak bases. The degree of dissociation pK_a is calculated from the following Henderson-Hassel Balch equations **11** and **12**.

In the case of acids:

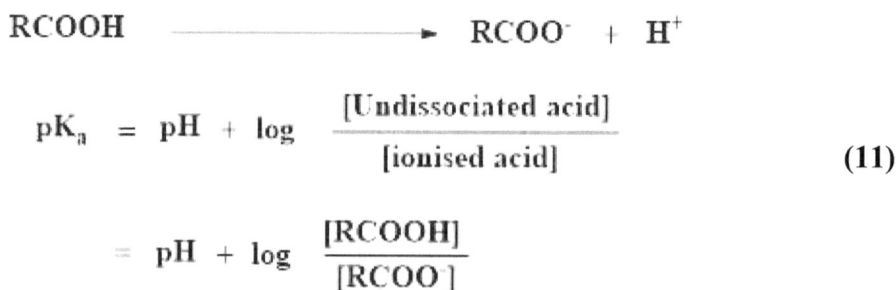

$$RCOOH \longrightarrow RCOO^- + H^+$$

$$pK_a = pH + \log \frac{[\text{Undissociated acid}]}{[\text{ionised acid}]} \tag{11}$$

$$= pH + \log \frac{[RCOOH]}{[RCOO^-]}$$

In case of bases;

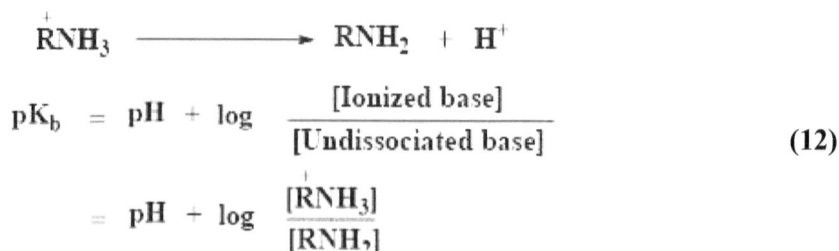

$$\overset{+}{R}NH_3 \longrightarrow RNH_2 + H^+$$

$$pK_b = pH + \log \frac{[\text{Ionized base}]}{[\text{Undissociated base}]} \tag{12}$$

$$= pH + \log \frac{[\overset{+}{R}NH_3]}{[RNH_2]}$$

Weak acids have a high pK_a; weak bases, a low pK_a.

Ionization

In general, drugs cross cellular membranes in undissociated form and act in dissociated form as ions. This happens because the passage of ions across the cellular membrane is facilitated by two factors:

a. The cellular membrane is made up of layers of electrically charged macromolecules (lipid, proteins, and mucopolysaccharides) that attract or repel ions.

b. Hydration of ions increases their volumes, rendering difficulty in their diffusion through pores.

Steric Parameters

Steric parameters or factors represent the form and the size of the substituent introduced in a parent molecule; that is, they measure the intermolecular steric effect. Examples of these parameters are Taft's substituent constants, Hancock's constant, van der Waals radii, steric constant, Charton's constant, molecular connectivity, and Verloop's sterimol parameters.

Toxicological Effect of Drugs on the Environment

LD$_{50}$ and ED$_{50}$

The quantity of a toxicant required to cause death is used as an indication of relative toxicity, which is most often expressed as LD_{50} (This expresses the mg/kg dose required to kill 50% of a test population). For example, the LD_{50} for sugar is >10,000 mg/kg, while for the botulism toxin it is ~ 0.0001 mg/kg. An alternative measurement of dose-response is ED_{50} where there is an effect on 50% of the test population [39, 40]. ED does not imply death. Pharmaceutical safety/effectiveness can be indicated by the pharmaceutical TI (therapeutic index). The therapeutic index is the ratio of LD_{50} / ED_{50}. A high TI value indicates an effective treatment at low doses and a lethal effect at higher doses. A low TI value indicates an ineffective treatment at low doses with lethality at the same low levels. The medicinal value of the drugs is generally represented by 'therapeutic index' or 'safety margin'. The therapeutic index is described as the ratio of the amount necessary to kill the patient [*i.e.* median lethal dose (LD_{50})] to that required for a median effective dose (ED_{50}). In experimental animals, the therapeutic index is calculated as:

$$TI = \frac{LD_{50}}{ED_{50}}$$

(13)

A therapeutic ratio of 10 indicates an LD_{50}: ED_{50} ratio of 10:1. This means that a ten-fold increase in the dose corresponding to ED_{50} would result in a 50% death rate.

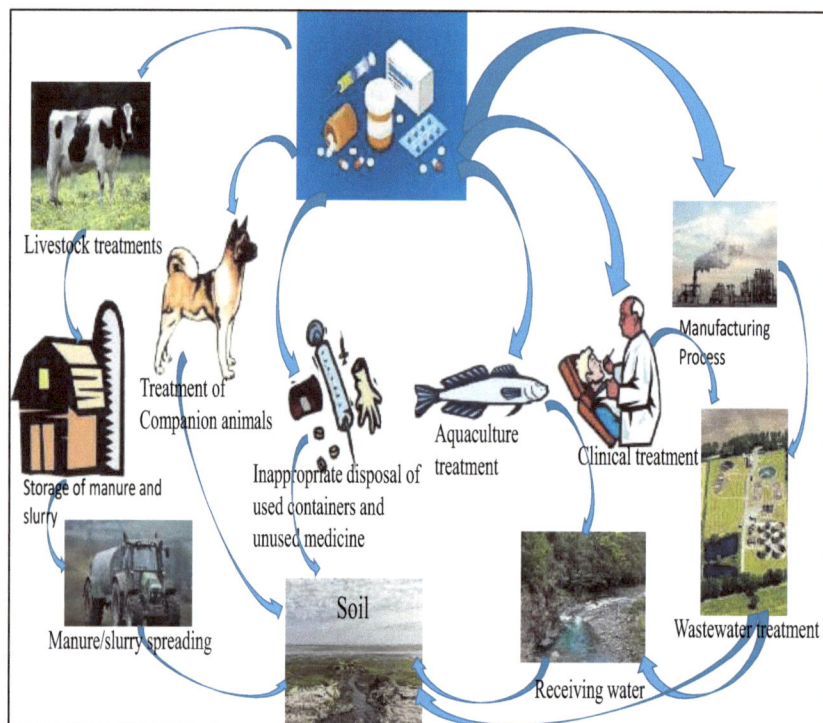

Fig. (20). Various routes of drugs entering the environment.

Effect of Drugs on the Environment

The presence of pharmaceutical drugs has created an eco-toxicity effect. The global consumption of pharmaceuticals is nearly 100,000 tonnes. During the manufacture of antibiotic, antimicrobial, generic medicines, active pharmaceutical ingredients, as well as other chemical ingredients, are directly released through sewage, streams, and many industrial effluents, into the environment. They are found in the concentration from sub ng/l to µg/l. For the treatment of sick farm animals, they are treated with veterinary medicines. These medicines are also used in intensive livestock treatments, and they indirectly enter the environment through the application of manure and slurry as fertilizers. Drugs used for the antibacterial treatment of aquatic fish or shrimp are directly released into surface water. The route by which drugs enter the environment is shown in Fig. (**20**) [41 - 43]. The anti-inflammatory drug diclofenac has shown dramatic pollution (>99%) leading to the extinction of gyps vulture species in India and Pakistan [41]. On demand, the manufacture of diclofenac has been doubly increased for the treatment of sick cattle; vultures suffered from renal failure after feeding on dead sick animals that were treated with diclofenac. The docking pose of the diclofenac drug with an amino acid of a living organism is shown in Fig. (**21**) [44, 45].

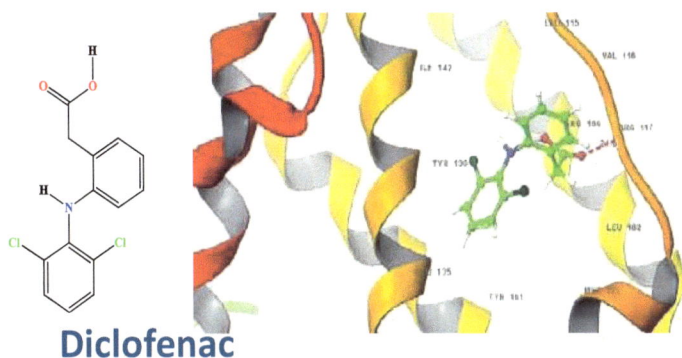

Fig. (21). Docking pose of diclofenac drug at active site with amino acid *via* hydrogen bonding.

CONCLUDING REMARKS

Drug design has improved the bioavailability of organic molecules as drugs. Organic molecules have been developed as drugs having a greater ability to cross the cell and membrane barriers, improved drug-targeting efficiency, and enhanced therapeutic effect. The physicochemical properties of the organic molecule receive much attention during the drug design, which depends on the size, functionality, solubility, and charge parameter. The drug modification has a great impact on its ability to cross and penetrate through the target biological barriers. This chapter provides vital insight into the drug design process that may be essential for the identification of therapeutic drug targets. In recent times, it has been successfully studied that residue of drugs could enter into the ecosystem through water channel and can, directly and indirectly, impact the soil, groundwater, and surface water, producing potential problems if not completely degraded. Therefore, drug design must also consider easy degradation of drug residues so that they do not contaminate the environment.

CONSENT FOR PUBLICATION

Not Applicable.

CONFLICT OF INTEREST

The author confirms that this chapter contents have no conflict of interest.

ACKNOWLEDGEMENT

Declared none.

REFERENCES

[1]   Schenone M, Dančík V, Wagner BK, Clemons PA. Target identification and mechanism of action in chemical biology and drug discovery. Nat Chem Biol 2013; 9(4): 232-40.
        [http://dx.doi.org/10.1038/nchembio.1199] [PMID: 23508189]

[2]   Singh S, Malik BK, Sharma DK. Molecular drug targets and structure based drug design: A holistic approach. Bioinformation 2006; 1(8): 314-20.
        [http://dx.doi.org/10.6026/97320630001314] [PMID: 17597912]

[3]   Sadiku MNO, Reeves SM, Musa SM. The impact of computational pharmacology. Eur Sci J 2019; 15: 151-6.
        [http://dx.doi.org/10.19044/esj.2019.v15n9p151]

[4]   Hodos RA, Kidd BA, Khader S, Readhead BP, Dudley JT. Computational approaches to drug repurposing and phamacology. Interdiscip Rev Syst Biol Med 2016; 8: 186-210.
        [http://dx.doi.org/10.1002/wsbm.1337] [PMID: 27080087]

[5]   Zhou SF, Zhong WZ. Drug design and discovery: Principles and applications. Molecules 2017; 22(2): 279.
        [http://dx.doi.org/10.3390/molecules22020279] [PMID: 28208821]

[6]   Upadhyay RK. Drug delivery systems, CNS protection, and the blood brain barrier. BioMed Res Int 2014; 2014869269
        [http://dx.doi.org/10.1155/2014/869269] [PMID: 25136634]

[7]   Doltra A, Dietrich T, Schneeweis C, *et al.* Magnetic resonance imaging of cardiovascular fibrosis and inflammation: from clinical practice to animal studies and back cardiovascular mri view project magnetic resonance imaging of cardiovascular fibrosis and inflammation: from clinical practice to Ani. BioMed Res Int 2013; 676489: 1-2.
        [http://dx.doi.org/10.1155/2013/676489]

[8]   Palleria C, Di Paolo A, Giofrè C, *et al.* Pharmacokinetic drug-drug interaction and their implication in clinical management. J Res Med Sci 2013; 18(7): 601-10.
        [PMID: 24516494]

[9]   Baig MH, Ahmad K, Rabbani G, Danishuddin M, Choi I. Computer aided drug design and its application to the development of potential drugs for neurodegenerative disorders. Curr. Neuropharmacol., 2017, 16, 740–748, (b) Najjar, A.; Najjar, A.; Karaman, R. Newly developed prodrug and prodrugs in development; An insight of the recent years. Molecules 2020; 25: 884-902.

[10]   Likhachev IV, Balabaev NK, Galzitskaya OV. Available instruments for analyzing molecular dynamics trajectoriesa. Open Biochem J 2016; 10: 1-11.
        [http://dx.doi.org/10.2174/1874091X01610010001] [PMID: 27053964]

[11]   Hefti FF. Requirements for a lead compound to become a clinical candidate. BMC Neurosci 2008; 9 (Suppl. 3): S7.
        [http://dx.doi.org/10.1186/1471-2202-9-S3-S7] [PMID: 19091004]

[12]   Hughes JP, Rees S, Kalindjian SB, Philpott KL. Principles of early drug discovery. Br J Pharmacol 2011; 162(6): 1239-49.
        [http://dx.doi.org/10.1111/j.1476-5381.2010.01127.x] [PMID: 21091654]

[13]   Pan W, Xue B, Yang C, *et al.* Biopharmaceutical characters and bioavailability improving strategies of ginsenosides. Fitoterapia 2018; 129: 272-82.
       [http://dx.doi.org/10.1016/j.fitote.2018.06.001] [PMID: 29883635]

[14]   aWen H, Jung H, Li X. Drug delivery approaches in addressing clinical pharmacology-related issues: Opportunities and challenges. AAPS J 2015; 17(6): 1327-40.
       [http://dx.doi.org/10.1208/s12248-015-9814-9] [PMID: 26276218] bKaraman R. Prodrugs design based on inter- and intramolecular chemical processes. Chem Biol Drug Des 2013; 82(6): 643-68.
       [http://dx.doi.org/10.1111/cbdd.12224] [PMID: 23998799]

[15]   Vale N, Ferreira A, Matos J, Fresco P, Gouveia M J. Amino Acids in the Development of Prodrugs 2018.
       [http://dx.doi.org/10.3390/molecules23092318]

[16]   Patil KD, Bagade SB, Sharma SR, Hatware KV. Potential of herbal constituents as new natural leads against helminthiasis: A neglected tropical disease. Asian Pac J Trop Med 2019; 12: 291-9.
       [http://dx.doi.org/10.4103/1995-7645.262072]

[17]   Hu Y, Lounkine E, Bajorath J. Many approved drugs have bioactive analogs with different target annotations. AAPS J 2014; 16(4): 847-59.
       [http://dx.doi.org/10.1208/s12248-014-9621-8] [PMID: 24871342]

[18]   Guha R. On exploring structure-activity relationships. Methods Mol Biol 2013; 993: 81-94.
       [http://dx.doi.org/10.1007/978-1-62703-342-8_6] [PMID: 23568465]

[19]   Cherkasov A, Muratov EN, Fourches D, *et al.* QSAR modeling: where have you been? Where are you going to? J Med Chem 2014; 57(12): 4977-5010.
       [http://dx.doi.org/10.1021/jm4004285] [PMID: 24351051]

[20]   Tungmunnithum D, Thongboonyou A, Pholboon A, Yangsabai A. Flavonoids and other phenolic compounds from medicinal plants for pharmaceutical and medical aspects: An overview. Medicines (Basel) 2018; 5(3): 93.
       [http://dx.doi.org/10.3390/medicines5030093] [PMID: 30149600]

[21]   Sykes DA, Charlton SJ. Slow receptor dissociation is not a key factor in the duration of action of inhaled long-acting β2-adrenoceptor agonists. Br J Pharmacol 2012; 165(8): 2672-83.
       [http://dx.doi.org/10.1111/j.1476-5381.2011.01639.x] [PMID: 21883146]

[22]   Conejo-Garcia A, Gallo MA, Espinosa A, Maria Campos J. 2D-QSAR: The mathematics behind the drug design methodology. New developments in med. Chem 2012; 1: 79-94.

[23]   Muratov EN, Bajorath J, Sheridan RP, *et al.* QSAR without borders. Chem Soc Rev 2020; 49(11): 3525-64.
       [http://dx.doi.org/10.1039/D0CS00098A] [PMID: 32356548]

[24]   Muratov EN, Artemenko AG, Varlamova EV, *et al.* Per aspera ad astra: application of Simplex QSAR approach in antiviral research. Future Med Chem 2010; 2(7): 1205-26.
       [http://dx.doi.org/10.4155/fmc.10.194] [PMID: 21426164]

[25]   Vekariya RL. A review of ionic liquids: Applications towards catalytic organic transformations. J Mol Liq 2017; 227: 44-60.
       [http://dx.doi.org/10.1016/j.molliq.2016.11.123]

[26]   Rhizobium GE. Complete genome sequence of the sesbania symbiont and rice. Nucleic Acids Res 2013; 1: 13-4.

[27]   Otero R, Vázquez de Parga AL, Gallego JM. Electronic, structural and chemical effects of charge-transfer at organic/inorganic interfaces. Surf Sci Rep 2017; 72: 105-45.
       [http://dx.doi.org/10.1016/j.surfrep.2017.03.001]

[28]   Bardo MT, Neisewander JL, Kelly TH. Individual differences and social influences on the neurobehavioral pharmacology of abused drugs. Pharmacol Rev 2013; 65(1): 255-90.

[http://dx.doi.org/10.1124/pr.111.005124] [PMID: 23343975]

[29]   McNerny DQ, Leroueil PR, Baker JR. Understanding specific and nonspecific toxicities: a requirement for the development of dendrimer-based pharmaceuticals. Wiley Interdiscip Rev Nanomed Nanobiotechnol 2010; 2(3): 249-59.
[http://dx.doi.org/10.1002/wnan.79] [PMID: 20166124]

[30]   Cascorbi I. Arzneimittelinteraktionen: Prinzipien, beispiele und klinische folgen. Dtsch Arztebl Int 2012; 109: 546-56.
[PMID: 23152742]

[31]   Stephenson RP. A modification of receptor theory. 1956. Br J Pharmacol 1997; 120(4) (Suppl.): 106-20.
[http://dx.doi.org/10.1111/j.1476-5381.1997.tb06784.x] [PMID: 9142399]

[32]   Salahudeen MS, Nishtala PS. An overview of pharmacodynamic modelling, ligand-binding approach and its application in clinical practice. Saudi Pharm J 2017; 25(2): 165-75.
[http://dx.doi.org/10.1016/j.jsps.2016.07.002] [PMID: 28344466]

[33]   Kenakin T. Efficacy at G-protein-coupled receptors. Nat Rev Drug Discov 2002; 1(2): 103-10.
[http://dx.doi.org/10.1038/nrd722] [PMID: 12120091]

[34]   Kenakin T. The mass action equation in pharmacology. Br J Clin Pharmacol 2016; 81(1): 41-51.
[http://dx.doi.org/10.1111/bcp.12810] [PMID: 26506455]

[35]   Liu Y, Guo W, Zhang Y, *et al.* Decreased resting-state interhemispheric functional connectivity correlated with neurocognitive deficits in drug-naive first-episode adolescent-onset schizophrenia. Int J Neuropsychopharmacol 2018; 21(1): 33-41.
[http://dx.doi.org/10.1093/ijnp/pyx095] [PMID: 29228204]

[36]   Grosman C, Auerbach A. The dissociation of acetylcholine from open nicotinic receptor channels. Proc Natl Acad Sci USA 2001; 98(24): 14102-7.
[http://dx.doi.org/10.1073/pnas.251402498] [PMID: 11717464]

[37]   Polishchuk PG, Muratov EN, Artemenko AG, Kolumbin OG, Muratov NN, Kuz'min VE. Application of random forest approach to QSAR prediction of aquatic toxicity. J Chem Inf Model 2009; 49(11): 2481-8.
[http://dx.doi.org/10.1021/ci900203n] [PMID: 19860412]

[38]   Omar NAS, Fen YW, Abdullah J, *et al.* Sensitive detection of dengue virus type 2 E-Proteins signals using self-assembled monolayers/reduced graphene oxide-PAMAM dendrimer thin film-SPR optical sensor. Sci Rep 2020; 10(1): 2374.
[http://dx.doi.org/10.1038/s41598-020-59388-3] [PMID: 32047209]

[39]   Patra JK, Das G, Fraceto LF, *et al.* Nano based drug delivery systems: recent developments and future prospects. J Nanobiotechnology 2018; 16(1): 71.
[http://dx.doi.org/10.1186/s12951-018-0392-8] [PMID: 30231877]

[40]   Raj J, Chandra M, Dogra TD, Pahuja M, Raina A. Determination of median lethal dose of combination of endosulfan and cypermethrin in wistar rat. Toxicol Int 2013; 20(1): 1-5.
[http://dx.doi.org/10.4103/0971-6580.111531] [PMID: 23833430]

[41]   Gunnarsson L, Snape JR, Verbruggen B, *et al.* Pharmacology beyond the patient - The environmental risks of human drugs. Environ Int 2019; 129: 320-32.
[http://dx.doi.org/10.1016/j.envint.2019.04.075] [PMID: 31150974]

[42]   Boxall ABA. The environmental side effects of medication. EMBO Rep 2004; 5(12): 1110-6.
[http://dx.doi.org/10.1038/sj.embor.7400307] [PMID: 15577922]

[43]   Weber FA, Bergmann A, Hickmann S, Ebert I, Hein A, Küster A. Pharmaceuticals in the environment–global occurrences and perspectives. Environ Toxicol Chem 2015; 35: 823-35.

[44]   Gan TJ. Diclofenac: an update on its mechanism of action and safety profile. Curr Med Res Opin

2010; 26(7): 1715-31.
[http://dx.doi.org/10.1185/03007995.2010.486301] [PMID: 20470236]

[45]   Nikalje AP, Hirani N. nawle, R.; Synthesis, *in-vitro*, *in-vivo* evaluation and molecular docking of 2-(3-(2-(1, 3-dioxoisoindolin-2-yl) acetamido)-4-oxo-2-substituted thiazolidin-5-yl) acetic acid derivatives as anti-inflammatory agents. Afr J Pharm Pharmacol 2015; 9(7): 209-22.
[http://dx.doi.org/10.5897/AJPP2014.4261]

CHAPTER 9

Advances in Biomolecular Simulations for Rational Drug Designing and Ecotoxicity

Viswajit Mulpuru[1] and Nidhi Mishra[1,*]

[1] Indian Institute of Information Technology Allahabad, Prayagraj, India

Abstract: This chapter emphasizes the advances in structure-based drug designing to accelerate the drug discovery process. This chapter discusses the various *in-silico* techniques, such as molecular docking, virtual screening, and molecular dynamics simulations, giving insight into quantum-chemical methods and quantitative structure-activity relationship (QSAR) techniques, which are some of the most popular methods in predicting drug efficiency that helps in designing novel molecular structures. It presents a clear concept of state-of-the-art computational techniques in molecular biology, pharmacology, and molecular medicine, using quantum-chemical techniques. Also, this chapter covers advances in environmental toxicity and its effect on human health. Pharmacological techniques, including pharmacokinetic and pharmacodynamic approaches, have been discussed to predict the effect of drugs on the environment and the human body, including the effects of toxic compounds on the environment and the human body. This chapter will be of immense value to readers of different backgrounds ranging from engineers and scientists to consultants and policymakers. It will be an invaluable resource for students, researchers, and industrial laboratories working in the areas related to medicinal chemistry, cheminformatics, pharmaceutical chemistry, pharmacoinformatic and environmental toxicology

Keywords: Drug design, Molecular docking, Molecular dynamics, PBPK, Pharmacodynamics, Pharmacokinetics, Physiology-based pharmacokinetics, QSAR, Quantitative Structure-Activity Relationship, Virtual screening.

INTRODUCTION

Bioinformatics, an interdisciplinary subject, has emerged as an essential discipline towards providing scientific and novel insights into various biological studies. It helps in better understanding the biomolecules through quantum and molecular-level analysis to complex systems biology approaches. Bioinformatics helps biologists to make use of the advances in computational technologies in various

* **Corresponding author Nidhi Mishra:** Indian Institute of Information Technology, Prayagraj, India; E-mail: nidhimishra@iiita.ac.in

Tahmeena Khan, Abdul Rahman Khan, Saman Raza, Iqbal Azad and Alfred J. Lawrence (Eds.)

biological applications, ranging from genomics and proteomics to advanced molecular dynamic simulations, systems biology, metabolomics, and pharmacokinetics. In this chapter, we have included various bioinformatic techniques such as molecular docking, virtual screening, molecular dynamics, QSAR, and pharmacokinetic modelling. As environmental pollution has proven to be a major hazard to the ecosystem, the applications of these bioinformatic techniques in the allied field of environmental sciences have been emphasized. These techniques help in understanding the effects of environmental pollutants on the human body and the ecosystem as a whole.

VIRTUAL SCREENING

Virtual screening is a technique to identify novel structural hits that show activities, such as inhibitory functions towards the specified proteins, from a large dataset of chemical libraries, using computational methods to identify drug candidates from a wide range of chemical libraries. The virtual screening technique employs similarity searching based on substructures, QSAR, and pharmacophore generation. Primarily, the virtual screening approach uses molecular docking and drug-likeliness analysis in finding the molecules that show biological activities towards the target. A schematic representation of various steps involved in virtual screening is shown in Fig. (**1**).

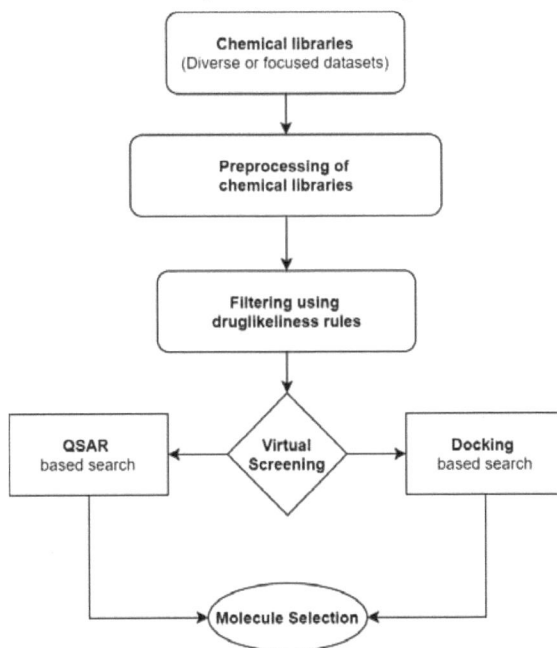

Fig. (1). A schematic representation of the steps involved in virtual screening.

MOLECULAR DOCKING

Molecular docking is a critical approach to achieve rational drug designing. It helps in understanding the molecular mechanisms involved in interactions between various biological and chemical structures, helping in analyzing their binding energies and affinities. It provides us with an overview of interactions between protein-ligand (Fig. **2**) or protein-protein complexes, ranking the candidate by their affinity scores. Molecular docking aims to evaluate the binding conformations of a ligand with a protein structure whose 3D conformation is known. The binding conformations of the ligand, known as binding pose, help in understanding the positioning and conformational state(s) of the ligand relative to the receptor. Therefore, docking methods can be used to rapidly screen a large number of small molecules into the binding site of a receptor, evaluating them in terms of binding energy between ligand and receptor.

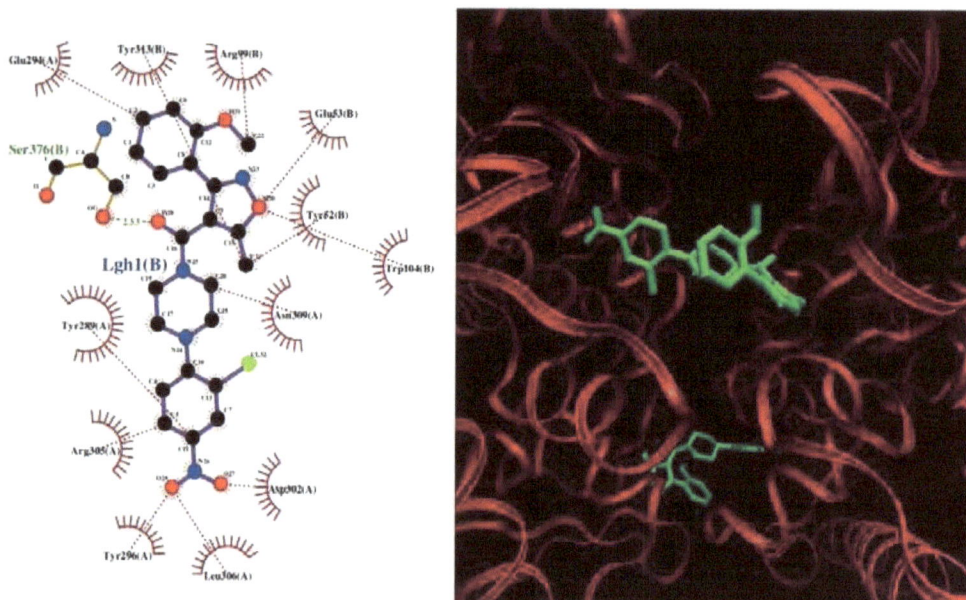

Fig. (2). Ligand bound to influenza A virus nucleoprotein.

Fig. (**2**) highlights the interactions between ligand and the nucleoprotein. The hydrogen bonds between the ligand and the protein are shown in green while the hydrophobic interactions are shown in red.

Docking can be used to predict the interactions between a ligand and a receptor (Fig. **2**). Generally, the ligand is a small biological molecule, but a peptide or a protein can also be used as a ligand [1]. The receptor is usually a protein of interest. Docking can be broadly classified into three types based on the flexibility of the receptor and the ligand [2, 3]. Higher the flexibility, higher the computational time needed to search for the best docking conformation.

Rigid Docking

Based on the lock and key model, in this type of docking, both the ligand and protein are considered rigid entities. This method is usually used for protein-protein docking, where the degrees of freedom to be sampled are very high.

Semi-Flexible Docking

One of the molecules, mainly the ligand, is flexible, while the protein is rigid, significantly reducing the degrees of freedom.

Flexible Docking

This is based on the concept of flexible protein of hand and glove model, also known as an induced fit model. In this method, both the protein and the ligand are considered flexible entities.

Before docking the ligand with a receptor, the binding site must be identified. This identification helps in minimizing the search space, thus reducing the computational time required to find the docking conformation [4]. Generally, the crystal structure of the protein is studied to identify the bound ligands whose binding location can be considered as the binding site of interest. In case the crystal structure does not contain a bound ligand, the binding site can be predicted using various tools, including freely available web servers. If the binding site is not identified, blind docking can be performed on the receptor by selecting the whole receptor as the search space, increasing the computational time.

After identifying the binding site, the ligand has to be docked onto the receptor at the recognized binding site. Various algorithms are available for docking the ligand with the receptor, ranging from semiempirical methods to knowledge-based approaches [5]. Machine learning-based approach is also being used for protein-ligand docking in recent time [6].

Being an important method for drug designing, molecular docking has various applications in drug discovery. It can be used to model new and efficient therapeutic compounds against various pathogens, including disease-causing bacteria and virus. It can also be used to identify DNA binding sites of different anticancer chemotherapeutic agents to study their interactions between the potential therapeutic agents and DNA. Further, the docking approach can be used in environmental studies to identify the effects and interactions of various pollutants towards living organisms.

ADMET ANALYSIS

The ADME (absorption, distribution, metabolism, excretion, and toxicity) analysis helps in studying the effects of a molecule on living organisms. This technique is majorly used in pharmaceutical industries to identify the drug likeliness of a compound towards an organism, especially humans. It can also be used to determine the effects of various environmental and industrial pollutants on humans and other living organisms.

The most common filter used to check the drug likeliness of a molecule is Lipinski's rule of five. This rule is very well-known and helps in identifying the permeability of the molecules in the human body. The filter detects the absorptivity of a molecule, considering four parameters. The molecule is predicted to show low absorptivity if the molecular weight (MW) is over 500, octanol/water partition coefficient (logP) is above 5, the molecule contains more than five hydrogen bond donors and has more than ten hydrogen bond acceptors

Veber rule, Ghose filter, and Blood-brain barrier rule are few other filters that are used in ADMET analysis.

MOLECULAR DYNAMICS

Molecular dynamics simulation is an important technique to study the dynamic nature of the biological molecules concerning their atomic and quantum level interactions. These atomic-level insights help explore the native structures and the stabilities of the biological molecules. Molecular dynamics simulation helps in unravelling many biological mysteries by simulating the natural conditions, ranging from a femtosecond to few milliseconds. Molecular dynamics simulations are widely used to study the stability of protein-ligand docking and protein-protein interactions [7]. Molecular dynamic simulations are also used for studying the effects of genomic and proteomic level mutations on these interactions and for protein fold prediction studies. In environmental studies, molecular dynamic simulations can be used for studying the stability and dynamic nature of various environmental pollutants and their effects on the human body and other organisms.

In molecular dynamics simulations, the molecules are usually modelled as a 'ball on spring' considering Hook's law. Generally, the trajectories of the atoms are determined using Newton's equations of motion, where forces between the particles and their potential energies are calculated with the help of various force fields. The most commonly used force fields in the biological simulation are AMBER, CHARMM and GROMOS, where the atomic representations are based on their nuclear radii [8 - 10].

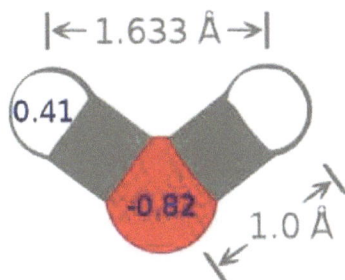

Fig. (3). Flexible SPC water model, the water model generally used during molecular dynamic simulations.

As stated earlier, the bonded interactions are represented by Hook's law while the nonbonded interactions are represented by Coulomb's and van der Waals interactions. Using Newton's laws, various algorithms have been developed for molecular dynamics simulations, which include the Verlet algorithm, velocity Verlet algorithm, Leapfrog algorithm, Langevin Dynamics and Maxwell-Boltzman Distributions [11].

General steps involved in Molecular Dynamics Simulation:

1. Energy minimization
2. Solvation
3. Heating and equilibration
4. The molecular dynamics production run
5. Analysis

Energy Minimization

The first step of molecular dynamic simulations is energy minimization. Energy minimization is performed for optimizing the geometry of the structure. Energy minimization is an essential step, as it is widely known and believed that molecules always tend to be in the lowest possible energy state for attaining the stable conformation. Although the conformational space can have a large number of local minima before reaching the global minima (Fig. **4**), energy minimization searches for the nearest local minima to perform the calculations in a finite time. Therefore, energy minimization depends on its initial conformation.

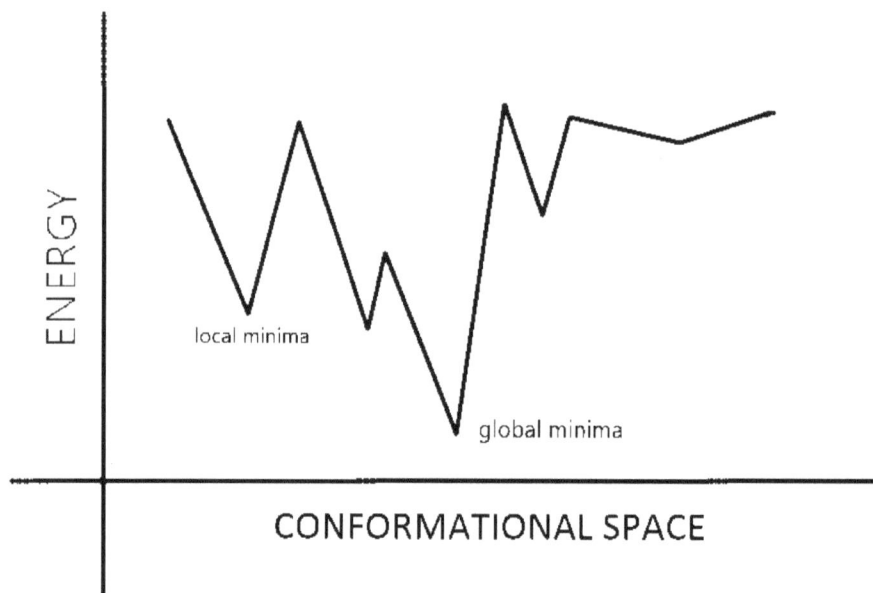

Fig. (4). Representation of the conformational space of a structure.

Solvation

As the molecular dynamic simulations are carried out in an environment mimicking the human body, *i.e.* the real environment, for studying the molecular dynamics in the human body it is essential to provide a cellular environment. As it is known that the cell is made up of seventy-five per cent of water and the remaining part consists of organic compounds such as carbohydrates, proteins and lipids, to reduce the complexity, only a water environment is provided during the simulation. To fill the boxed environment with water molecules, various implicit water models are available, such as SPC (Fig. **3**), SPC/E, TIP3P, TIP4P, and TIP5P [12]. Different organic and lipid-based solvent models are also available for mimicking various biological environments.

Heating, Equilibration and Production Run

To mimic the native environment, the system generated model has to be equilibrated at the required temperature before productions run. When considering the human body, the temperature is set at 310 K. The temperature is increased linearly from 0 to 300 K using Maxwell Boltzmann's distribution [11]. After attaining the required temperature, the system is equilibrated to ensure its stability. Once the system is equilibrated, production runs are invoked to study the dynamic nature of the system. The production runs can range from femtoseconds to a few milliseconds, depending on the study.

Analysis

After incorporating the production runs, they are analysed to retrieve required information from molecular dynamics simulation to study their dynamic nature and atomic-level interactions. A few analysis methods are given below:

Structural Visualization

Various visualization tools like VMD [13] can be used to visualize the simulation results to study the quantum level changes in the structural complex, including interactions between protein-protein, protein-ligand and protein-nucleic acids.

Time Series and Thermodynamic Calculation

Various time series calculations like root-mean-square deviation (RMSD), root mean square fluctuation (RMSF), timeline analysis of secondary structure evolution, change in nonbonded interaction energy *etc.* (Fig. **5**) can be performed to evaluate the atomic and molecular level changes in the system over the simulation time. RMSD and RMSF help in understanding the stability and flexibility of the system and residues, respectively. In contrast, the secondary structure evolution helps in understanding the stability of the secondary structure of the protein. The nonbonded interaction energy between the protein and ligand helps in understanding the binding stability of the molecules.

To further study the dynamic nature of the complex, thermodynamics calculations can be incorporated into the study. The thermodynamic calculations usually include bonded and nonbonded interactions, such as covalent, electrostatic, and hydrogen bonds and van der Waals interactions. The Molecular Mechanics with Poisson Boltzmann and Solvent Accessibility (MM-PBSA) and Molecular Mechanics with Generalized Born and Solvent Accessibility (MM-GBSA) can also be used to calculate binding energies by combining molecular dynamics and free energy scoring schemes [14].

Molecular dynamics simulations can be applied in various fields, ranging from film industries to scientific studies like geology, physics, biology, and environmental sciences [15, 16]. In biology, it can be used in multiple studies, ranging from analyzing changes in quantum level molecular interactions and molecular docking to protein folding. As stated earlier, molecular dynamics simulations can be of great help in environmental studies for analyzing the effects of ecological factors and industrial pollutants on the human body.

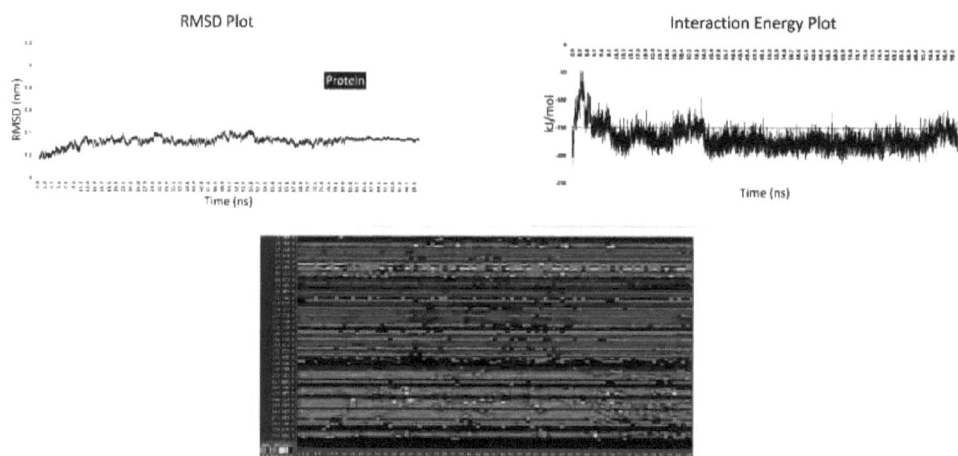

Fig. (5). The RMSD plot for molecular dynamics simulation of a protein simulation for 100ns (Top left), the interaction energy plot between a small molecule and a protein during molecular dynamic simulation of 100ns (Top right), the secondary structure evolution timeline of a protein during a molecular dynamic simulation (Bottom).

Quantitative Structure-Activity Relationship (QSAR)

The quantitative structure-activity relationship (QSAR) is a part of physical organic chemistry that generally helps in understanding the toxicology of the molecules. QSAR helps in reducing the time and money required for drug discovery, decreasing the number of molecules needed to synthesize and screen. QSAR techniques can be applied to various fields, including materials science, environmental sciences and nanotechnology for predicting the activity, property, and toxicity of the compounds.

QSAR identifies and quantifies various physicochemical properties of the compounds to study their role on the molecule's biological activity or its toxicity. An equation is generated using the parameters that affect the required property of the molecule, suggesting the pharmacokinetic importance of these parameters. These parameters can include the structural, physical, or chemical properties of the molecule. Although molecules have a large number of parameters, it is usually appropriate to consider only a few parameters to reduce the time taken for calculations as it could be tedious to quantify all the parameters to the activity. Generally, one or two physicochemical properties are considered to generate a QSAR model. Fig. (7) gives the schematic representation of the steps involved in QSAR model building and analysis.

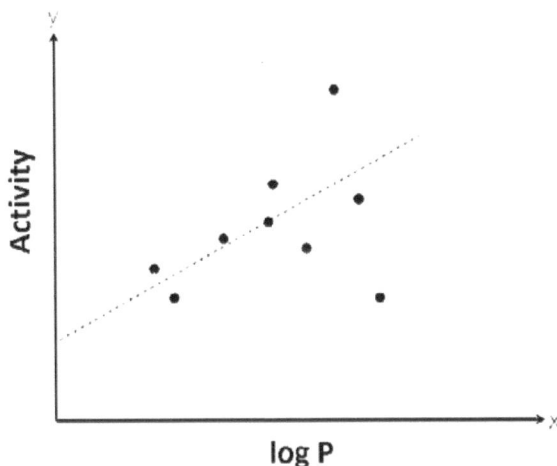

Fig. (6). Biological activity *versus* log P.

To generate a simple QSAR model, a single parameter can be correlated with the biological activity, thus generating a simple graph using the linear regression method where the line closest to the data points is considered as the best. In the linear regression method, the sum of squares of the distance between the points and the line is calculated. The best line will have the minimum sum of squares.

The best line is calculated using the straight-line equation $y = ax + b$, where a and b are the constants that are varied to find the best fit (Fig. **6**). Further, the regression coefficient (r) can be calculated on the best fit, which states the accuracy of the fit. The closer the value of r is to 1, the better is the accuracy of the fit. Various other statistical measures like the standard error of estimate (standard deviation) and F-tests can be performed on the fit to check the accuracy of the model further.

Physiochemical Parameters

The parameters that are used for generating the descriptors of the molecule range from physical and structural to chemical properties. The commonly used parameters are the quantifiable hydrophobic, electronic, and steric properties.

Hydrophobicity

The hydrophobicity of the molecule helps in studying the ability of the molecule in crossing the cell membrane, hence affecting the biological activity. The hydrophobicity of the molecule can be estimated employing the partition coefficient (P) of the molecule and the hydrophobicity constant of the substituent (π).

The Partition Coefficient (P)

The hydrophobicity of a molecule can be measured by measuring the relative distribution of the molecule in an n-octanol/water mixture, as shown in equation **1**. The Hydrophobic molecules will prefer the octanol layer while the hydrophilic molecules will prefer the water layer. This relative distribution is the partition coefficient denoted by *p*.

$$p = \frac{\text{The concentration of drug in n-octanol}}{\text{The concentration of drug in water}} \qquad (1)$$

The Substituent Hydrophobicity Constant (π)

When a large number of molecules are to be tested for their hydrophobicity, it will be a tedious and time-intensive process. To decrease the time required for calculating the hydrophobicity of the molecule, the hydrophobicity of the new molecule can be obtained by summating the partition coefficient of the individual substituents present in the required molecule, as seen in equation **2.**

$$log\ P_{benzamide} = log\ P_{benzene} + \pi_{amide} \qquad (2)$$

To obtain the hydrophobic constant of the individual substituent, the partition coefficient is measured experimentally for a standard molecule with and without a substituent and the partition coefficient of the standard molecule is subtracted from the partition coefficient of the standard molecule with a substituent (Eq. **3**). The hydrophobicity constant of the substituent is denoted by π.

$$\pi_x = log\ P_x - log\ P_s \qquad (3)$$

where P_x is the partition coefficient of the standard containing substituent while $log\ P_s$ is the partition coefficient of the standard.

Electronic Effects

When the molecule interacts with various complexes in the body and the environment, it shows various electronic effects by establishing bonds. These bonds are formed by disruptive forces and show charge transfer complex formation, ionic interaction, inductive effects, hydrogen bonds, polarization effect, and catalytic property characterize electronic features. Again, some of these parameters employ multiple mechanistic features. The electronic effects are generally calculated utilizing the Hammett substituent constant, denoted by σ, and the dissociation constant of the molecule [17].

Steric Factors

Steric features are related to the 2D structure of the molecule. The steric factors are essential for the molecule to show specific interaction with various environmental complexes, such as proteins at the suitable spatial space. The steric factors show influence in the contribution of substituents in interactions, molecular size and shape, the conformation of the molecules, and the spatial arrangement of the molecule. There are various methods to calculate the steric factors of the molecule [17] such as Taft's steric factor (E_s), Molar refractivity (MR), and Verloop steric parameters.

Other molecular parameters, such as van der Waals interaction, dipole moments, hydrogen bonding, and interatomic distances, are also used in generating a QSAR model. The basic workflow of QSAR analysis is depicted in Fig. (**1**).

Hansch Equation

As we have studied, QSAR models can be generated using one molecular property by drawing up a simple equation. However, generally, this is not the case with many of the molecules. For most of the molecules, the equation has to be drawn up using a combination of various physicochemical parameters. In situations like these, drawing up a simple equation is tough. In such cases, the models are generated by incorporating various parameters into a single model. The equations generated for these models are known as Hansch equations, which involve various physicochemical properties, such as log P, π, σ, and steric factors [18]. The general Hansch equation is seen in equation **4**:

$$log \left(\frac{1}{c}\right) = a \, logP + b \, \sigma + c \, E_s + d \tag{4}$$

where a, b, c, and d are constants that are determined during the generation of the equation.

As the behaviour of the molecules largely depends on the structural and physicochemical parameters, QSAR analysis helps in developing a rational strategy toward drug designing. A significant change in pharmacological/toxicological activity is shown even with a minimal change in the structure. QSAR reduces the time required for practical experimentation by using the determining conclusions from theoretical studies. QSAR helps pharmacologists by assessing the biological activities of the molecules. As stated earlier, QSAR helps in determining the toxicity of the molecules towards the human body and environment, helping the environmentalists access the effects of industrial pollutants.

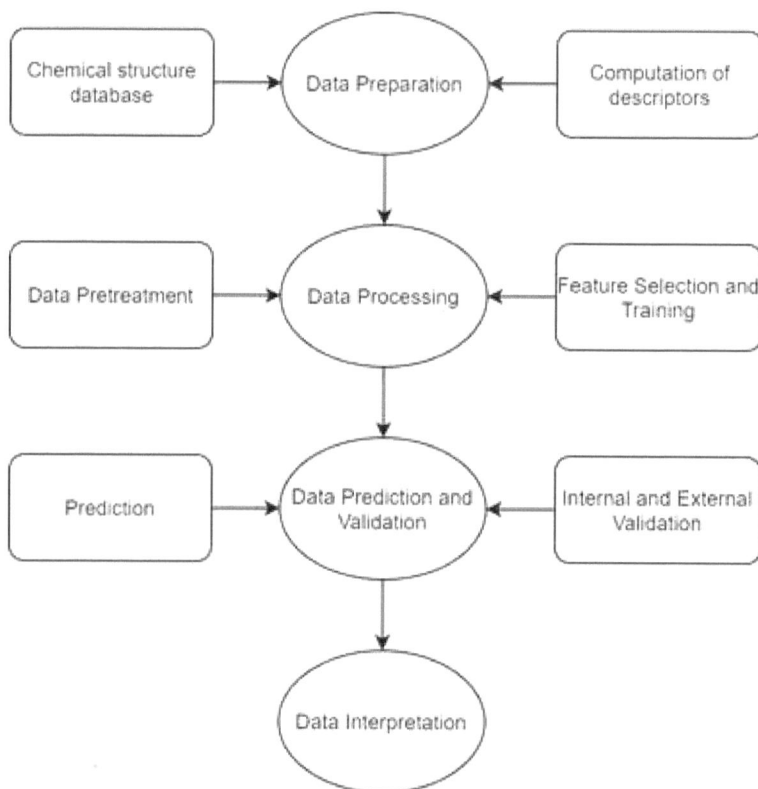

Fig. (7). A schematic representation of the methodology involved in QSAR analysis.

Pharmacokinetics and Pharmacodynamics

Pharmacodynamics is the study of molecules and their interaction with molecular targets that show pharmacological activity such as proteins and enzymes. Pharmacokinetics helps in studying the effects and activities of the molecules on the whole organism. It is an essential step in identifying the toxicity and side effects of the molecules using theoretical methods, thereby reducing the time to find suitable candidates with the required activity before the clinical trial stage [19].

Until now, we have studied drug designing, binding interactions and targeting, which are the pharmacodynamic aspects of a molecule. But before the molecule could start showing activity, it should reach its prescribed target. So, it is a crucial step to study the pharmacokinetics of the molecule along with its pharmacodynamics. As stated, pharmacokinetics consists of four primary steps absorption, distribution, metabolism, and excretion (ADME). Pharmacokinetics is an essential technique in studying the effects of environmental and industrial pollutants on living organisms, including humans, and their absorptivity and toxicity to the organisms.

Absorption

For a molecule to get absorbed into the body *via* the oral route, it has to contain a correct balance of water and lipids. If the molecule is highly hydrophilic, it might fail to pass the lipid membranes of the gastrointestinal tract. In case the molecule is hydrophobic, it might dissolve in the globular cells without reaching the specified target. In case the molecule gets deposited in globular cells of the body, it might not leave the body and over time, the concentration might keep increasing, causing unwanted toxicity and side effects.

Molecular absorption through the skin is also a significant concern as the atmospheric toxins might get readily absorbed in the bloodstream, in case the toxin is made up of the right amount of hydrophobicity and molecular weight to penetrate the layers of the skin. The pH and pK also play a significant role in the absorption of the molecule. The absorptivity of the molecule can be determined to study the availability of its ionized form at a particular pH. The ionization extent of a molecule at a given pH can be determined using the dissociation constant, by the Henderson–Hasselbalch equation (Eq. **5**).

$$pH = pK_a + \log_{10} \frac{\text{Concentration of the unionized molecule}}{\text{Concentration of the ionized molecule}} \tag{5}$$

Although molecular weight plays a significant role in the absorptivity of the molecule, a molecule having high molecular weight might pass through the membranes by pinocytosis. In pinocytosis, the molecule is engulfed by the cell membrane and then released onto the other side of the membrane. Pinocytosis largely depends on the hydrophobicity of the molecule and membrane. Considering the water percentage of the body, usually highly polar molecules show this kind of absorptivity. Various other types of molecular absorptivity include oral absorption, absorption through the mucous membrane, inhalation, and injection.

Distribution

Once a molecule gets absorbed, it gets rapidly distributed along with the blood supply and slowly enters various tissues and organs. The rate of distribution of the molecule depends on various physicochemical factors of the molecule and the body. Once the drug distributes evenly in the blood, it starts entering the tissues and cells to show its activity. This distribution of the molecule can be studied to find the targets of the given molecule. In drug designing, the drugs are tuned for optimal drug targeting. Once the molecule enters the tissues and organs, its concentration in the blood starts decreasing.

Unlike various organs of the body, the brain is lined with tightly packed cells without any pores. If a molecule has to enter the blood, it has to negotiate the blood-brain barrier. As this barrier is made up of a fatty layer, the polar molecules cannot pass through this barrier. As the placental barrier that separates the mother and fetal blood majorly consists of fatty cells, fat-soluble molecules including alcohol, cocaine and nicotine, pass through this barrier, affecting the fetus.

Metabolism

As we all know, as soon as a molecule enters the bloodstream, it is subjected to various metabolic enzymes. These enzymes help in degrading or metabolizing the molecule, to be easily excreted. During metabolism, the molecules might form various intermediate metabolites, which might be toxic to the organism. Therefore, studying the metabolic reaction of the molecule is very important to estimate the toxicity of the molecule towards the organisms. In drug designing, the side effects of the drug can be predicted by metabolic reactions.

Various enzymes that help in metabolizing the molecule as it enters the blood are cytochrome P450 enzymes, flavin-containing monooxygenases from the liver and various other oxidative and hydrolytic enzymes, such as monoamine oxidases and esterases, from different tissues of the body.

Excretion

Once, the molecules are metabolized, they are excreted out of the body by various routes. Volatile and gaseous molecules, including gaseous anaesthetics, are excreted out through the lungs. Other molecules that are metabolized pass along the bile from the liver to the intestine for excretion. Once the molecule reaches the intestine, it has a chance to get reabsorbed into the body through the intestinal walls. Other routes of molecular excretion include elimination through sweat, saliva, breast milk, and kidney.

Pharmacokinetic Models

Pharmacokinetic models help in studying the ultimate fate of molecules that have entered a biological system. These models are generated taking into consideration various pharmacokinetic parameters, namely absorption, distribution, metabolism and excretion, as studied [20]. Three major types of pharmacokinetic models are:

   i. One-compartment model
  ii. Two-compartment model
 iii. Multicompartment model

Various basic parameters that are used to generate a pharmacokinetic model are:

The elimination rate constant

This states the rate at which the molecule is eliminated from the body with respect to time. The rate of elimination can be described as shown in equation **6**:

$$X = X_0 \, exp(-kt) \tag{6}$$

where X is the amount of drug, X_0 is the initial dosage of the drug and k is the first-order elimination rate constant.

The volume of distribution

The volume of distribution (V_d) states the volume of blood in which the given drug is distributed. Although the concentration of the molecule inside tissue is not the same as the concentration of the molecule in the blood, their concentrations remain proportionate to each other. Thus the volume of the distribution corresponds to the concentration of the molecule in the blood. The V_d can be calculated using equation **7**.

$$V_d = \frac{x}{c_b} \tag{7}$$

where Cb is the concentration of the molecule in the blood.

Half-Life

This parameter helps in the estimation of the time required for the molecule to reduce to half of its initial concentration in blood. This helps in estimating the time interval required between doses to prevent the molecule from attaining toxic concentrations. For a linear one-compartment model, the half-life is calculated as shown by equation **8**.

$$t_{1/2} = \frac{0.693}{k} \tag{8}$$

where $t_{1/2}$ is the half-life in sec (s) and **k** is the rate constant in sec^{-1} (s^{-1}).

Clearance

Clearance (CL) signifies the volume of blood cleared per unit time, either by metabolism or by excretion. Considering the activities of the kidney and liver, the total clearance of the molecule from the body is determined by the following formula (Eq. **9**).

$$CL = CL_{renal} + CL_{nonrenal} \tag{9}$$

Considering the elimination rate constant and volume of distribution

$$CL = k * V_d \tag{10}$$

Hence the half-life of the molecule can be determined by equation **11.**

$$t_{1/2} = \frac{0.693 \: X \: V_d}{CL} \tag{11}$$

Pharmacokinetic modelling with the help of various parameters helps in understanding the pharmacokinetic effects of various molecules on the human body and other living organisms. It helps in understanding the dosage and time interval of the drugs required to show activity and reduce toxicity. In environmental studies, pharmacokinetic models help in understanding the effects of various environmentally active molecules on organisms. It helps in assessing the safety and toxicity of the pollutant at various concentrations on the whole body.

Applications of Bioinformatic Techniques in Environmental Sciences

Environmental pollutants are a significant hazard to the world. Although industrialization and urbanization are considered as a benchmark of development, the environmental pollution caused by these activities affects the lives of all kinds of organisms, from bacteria, marine life, and birds to humans, leading to disturbance in the ecological balance. Bioinformatics may play an important role in mitigating the effects of these activities by helping in studying the effects of pollutants on the environment as well as in developing various aspects of bioremediation.

The techniques used in virtual screening, like molecular docking and toxicity analysis, can help in understanding the pharmacodynamic effect of the pollutants on living organisms. A simple case study on this approach could consist of identifying an industrial pollutant, causing significant water pollution, and studying its effects on various proteins of the human body. This approach employs binding affinity analysis to study the effect of this pollutant as an inhibitor of various enzymes by blocking their active sites, thus causing disorders.

The Molecular Dynamics approach is a vital tool in environmental studies. This technique helps in simulating a natural environment and understanding the effect of various environmental factors and pollutants on organisms. The molecular dynamic simulations can range from a simplistic water model for understanding the effects of pollutants on a protein to creating a single or multicellular cell model constituting various organelles, thus helping in understanding the effects of unwanted industrial and environmental compounds on the cell, with significant accuracy.

For example, phenols are major environmental pollutants that are not only synthesized industrially but also produced by living organisms, such as plants and microbes that get accumulated in surface and groundwater, including soil when released into the environment [21 - 23]. The major sources of phenols in industrial wastewaters include oil, textile, and pharmaceutical industries. It is known that phenol and its vapours are harmful to the respiratory tract and the skin due to their protein denaturation properties [24, 25].

Studies on phenols suggest that docking techniques can be adapted to study the interactions between the enzymes and phenols at the molecular level in the process of biodegradation. It is observed that the phenols form hydrogen bonds with laccase enzymes that aid in oxidation reactions at the binding pocket [26]. Similarly, the binding locations of phenolic compounds on human proteins can be predicted by considering important protein complexes from the respiratory tract, skin, and eyes, that are affected by phenolic pollutants to study their mechanism of action on the human body. Further, the molecular dynamics simulation studies can be incorporated to study the interaction stability of these pollutants with the human proteins in a simple water box or a more complex lipid-based *in-silico* cell-like environment.

The QSAR technique can be used to generate various models of pollutants. These models can further be used to understand the effects of new pollutants on organisms and the ecological balance. This technique significantly reduces the time required to deduce the effects of pollutants on the environment.

Pharmacokinetic modelling helps in generating accurate models of ecological aspects, ranging from a single organism to a complete ecosystem. These models can help in understanding the effect of pollutants on the whole ecosystem starting from a fundamental level. A significant case study through this approach could include generating a simple ADME model of a unicellular organism and studying the effect of a pollutant on this organism, from absorption and metabolism to excretion. The same model could further be expanded to include various organisms and studying the effects of these excreted compounds on other species, thus including a whole ecosystem and studying the effects of pollutants on the whole of it.

The pharmacokinetic models can further help in understanding the effects of various organisms in bioremediation. The metabolic activity of different organisms can be modelled with the help of pharmacokinetics, and these models can be subjected to various environmental and industrial pollutants to understand the bioremediation activity of these organisms on these pollutants.

For example, physiologically based pharmacokinetic (PBPK) models of humans are made up of compartments corresponding to the different organs of the body connected by the cardiovascular system. The compartments are described by physical characteristics, such as tissue volume, blood, and fluid flow rate specific to the specific organ. The tissues are defined by perfusion-rate and permeability-rate where perfusion-rate kinetics occur for small lipophilic molecules while permeability-rate kinetics exist for larger and hydrophilic molecules that permeate across the cell membrane. These perfusion-rate and permeability-rate help in understanding the absorption rate of the cell, organ, and organism. The PBPK models of intestine and liver are modelled to predict the contributions of enzymes and transporters on the metabolism of an organism considering enzymes, such as P450 isoenzyme and glutathione transferase [27, 28].

Similarly, the PBPK model of a unicellular organism can be modelled to predict its absorptivity and metabolic rate of an environmental pollutant. These pharmacokinetic models, along with metabolic pathway analysis, can help in understanding the role of microbes in bioremediation and also in engineering microbes and other organisms to perform and enhance the rate of bioremediation through metabolic pathway engineering.

CONCLUDING REMARKS

This chapter discusses different bioinformatic techniques useful in various studies of biotechnology, life science, environmental sciences, and other allied subjects of biosciences. This chapter introduces the basics of various techniques, including molecular docking, virtual screening, molecular dynamics, QSAR, and pharmacokinetic modelling, to the reader. Molecular docking helps in analyzing the physiochemical changes in a molecule and its interactions with small molecules, such as inhibitors. While docking commonly considers a static structure of a system, the molecular dynamic simulation gives a picture of the dynamic nature of the system. This technique is used in simulating various biological compounds, such as proteins, nucleic acids, and carbohydrates, to understand their interaction stability with small biologically significant molecules. It helps in solving the biological mysteries that occur in a biological system by studying the dynamicity of quantum level and molecular interaction, within a few milliseconds. Various methods and steps of docking and molecular dynamic simulation, including the virtual screening approach, were discussed in the chapter. The QSAR technique attempts to identify and quantify the physicochemical properties of a molecule to understand the effects of these properties on biological activity. In case the physicochemical properties affect the biological activity, a model is generated to quantify the role of this property in the pharmacokinetics of the molecule. Although pharmacodynamics plays an essential role in drug designing, the study of pharmacokinetics is an essential step in understanding the effect of the drug on the whole organism. Pharmacokinetics helps in building a theoretical model of organisms and environment to study the effects of different molecules on the whole system by employing and analyzing the absorption, distribution, metabolism, and excretion parameters of the system. Various properties, parameters, and equations required to generate the QSAR and pharmacokinetic models were discussed in the chapter. As bioinformatics techniques are an emerging and promising approach towards environmental studies to help understand and contain environmental degradation, the applications of bioinformatic techniques towards environmental sciences have been discussed in this chapter. Overall, the chapter explores the basic concepts of bioinformatics and their uses in its allied disciplines, such as environmental studies, as the interconnection among disciplines is crucial to mitigate the scientific barriers towards sustainable development.

CONSENT FOR PUBLICATION

Not Applicable.

CONFLICT OF INTEREST

The author confirms that this chapter contents have no conflict of interest.

ACKNOWLEDGEMENT

Declared none.

REFERENCES

[1]   Vakser IA. Protein-protein docking: From interaction to interactome. Biophys J 2014; 107(8): 1785-93.
        [http://dx.doi.org/10.1016/j.bpj.2014.08.033] [PMID: 25418159]

[2]   Meng X-Y, Zhang H-X, Mezei M, Cui M. Molecular docking: a powerful approach for structure-based drug discovery. Curr Comput Aided Drug Des 2011; 7(2): 146-57.
        [http://dx.doi.org/10.2174/157340911795677602] [PMID: 21534921]

[3]   Hernnndez-Santoyo A, Yair A, Altuzar V, Vivanco-Cid H, Mendoza-Barrer C. Protein-Protein and Protein-Ligand Docking. 2013.
        [http://dx.doi.org/10.5772/56376]

[4]   Laurie AT, Jackson RM. Methods for the prediction of protein-ligand binding sites for structure-based drug design and virtual ligand screening. Curr Protein Pept Sci 2006; 7(5): 395-406.
        [http://dx.doi.org/10.2174/138920306778559386] [PMID: 17073692]

[5]   Novič M, Tibaut T, Anderluh M, Borišek J, Tomašič T. The comparison of docking search algorithms and scoring functions.Methods and Algorithms for Molecular Docking-Based Drug Design and Discovery. IGI Global 2016; pp. 99-127.
        [http://dx.doi.org/10.4018/978-1-5225-0115-2.ch004]

[6]   Shen C, Ding J, Wang Z, Cao D, Ding X, Hou T. From machine learning to deep learning: Advances in scoring functions for protein–ligand docking. Wiley Interdiscip Rev Comput Mol Sci 2020; 10e1429
        [http://dx.doi.org/10.1002/wcms.1429]

[7]   Karplus M, McCammon JA. Molecular dynamics simulations of biomolecules. Nat Struct Biol 2002; 9(9): 646-52.
        [http://dx.doi.org/10.1038/nsb0902-646] [PMID: 12198485]

[8]   Wang J, Wolf RM, Caldwell JW, Kollman PA, Case DA. Development and testing of a general amber force field. J Comput Chem 2004; 25(9): 1157-74.
        [http://dx.doi.org/10.1002/jcc.20035] [PMID: 15116359]

[9]   Vanommeslaeghe K, Hatcher E, Acharya C, *et al.* CHARMM general force field: A force field for drug-like molecules compatible with the CHARMM all-atom additive biological force fields. J Comput Chem 2010; 31(4): 671-90.
        [PMID: 19575467]

[10]   Oostenbrink C, Villa A, Mark AE, van Gunsteren WF. A biomolecular force field based on the free enthalpy of hydration and solvation: the GROMOS force-field parameter sets 53A5 and 53A6. J Comput Chem 2004; 25(13): 1656-76.
        [http://dx.doi.org/10.1002/jcc.20090] [PMID: 15264259]

[11]   Toxvaerd S. Molecular dynamics at constant temperature and pressure. Phys Rev E Stat Phys Plasmas Fluids Relat Interdiscip Topics 1993; 47(1): 343-50.
        [http://dx.doi.org/10.1103/PhysRevE.47.343] [PMID: 9960009]

[12]   Chevrot G, Schurhammer R, Wipff G. Molecular dynamics simulations of the aqueous interface with the [BMI][PF6] ionic liquid: Comparison of different solvent models. Phys Chem Chem Phys 2006; 8(36): 4166-74.
        [http://dx.doi.org/10.1039/b608218a] [PMID: 16971984]

[13]   Humphrey W, Dalke A, Schulten K. VMD: visual molecular dynamics. J Mol Graph 1996; 14(1): 33-38, 27-28.

[http://dx.doi.org/10.1016/0263-7855(96)00018-5] [PMID: 8744570]

[14]   Genheden S, Ryde U. The MM/PBSA and MM/GBSA methods to estimate ligand-binding affinities. Expert Opin Drug Discov 2015; 10(5): 449-61.
[http://dx.doi.org/10.1517/17460441.2015.1032936] [PMID: 25835573]

[15]   Ma Z, Gamage RP, Rathnaweera T, Kong L. Review of application of molecular dynamic simulations in geological high-level radioactive waste disposal. Appl Clay Sci 2019; 168: 436-49.
[http://dx.doi.org/10.1016/j.clay.2018.11.018]

[16]   Karplus M, Petsko GA. Molecular dynamics simulations in biology. Nature 1990; 347(6294): 631-9.
[http://dx.doi.org/10.1038/347631a0] [PMID: 2215695]

[17]   Hansch C, Leo A, Taft R. A survey of Hammett substituent constants and resonance and field parameters. Chem Rev 1991; 91: 165-95.
[http://dx.doi.org/10.1021/cr00002a004]

[18]   Fujita T. The extrathermodynamic structure-activity correlations. Advances in Chemistry. American Chemical Society 1974; pp. 1-19.

[19]   Patrick GL. Pharmacodynamics and pharmacokinetics.An Introduction to Medicinal Chemistry. 5th ed. Oxford University Press 2013; pp. 87-153.

[20]   Dhillon S, Gill K. Basic pharmacokinetics.Clinical Pharmacokinetics. London: Pharmaceutical Press 2006; pp. 1-44.

[21]   Hättenschwiler S, Vitousek PM. The role of polyphenols in terrestrial ecosystem nutrient cycling. Trends Ecol Evol 2000; 15(6): 238-43.
[http://dx.doi.org/10.1016/S0169-5347(00)01861-9] [PMID: 10802549]

[22]   Khoddami A, Wilkes MA, Roberts TH. Techniques for analysis of plant phenolic compounds. Molecules 2013; 18(2): 2328-75.
[http://dx.doi.org/10.3390/molecules18022328] [PMID: 23429347]

[23]   Gianfreda L, Sannino F, Rao MA, Bollag JM. Oxidative transformation of phenols in aqueous mixtures. Water Res 2003; 37(13): 3205-15.
[http://dx.doi.org/10.1016/S0043-1354(03)00154-4] [PMID: 14509708]

[24]   Ullmann's Encyclopedia of Industrial Chemistry. 2011.

[25]   The merck index: An encyclopedia of chemicals, drugs, and biologicals, 15th ed.; O'Neil, M. J., Heckelman, P. E., Dobbelaar, P. H., Roman, K. J., Kenny, C. M., Karaffa, L. S., Royal Society of Chemistry (Great Britain), Eds.; Royal Society of Chemistry: Cambridge, UK, 2013.

[26]   Zhang Y, Zeng Z, Zeng G, *et al.* Effect of Triton X-100 on the removal of aqueous phenol by laccase analyzed with a combined approach of experiments and molecular docking. Colloids Surf B Biointerfaces 2012; 97: 7-12.
[http://dx.doi.org/10.1016/j.colsurfb.2012.04.001] [PMID: 22580478]

[27]   Cuello WS, Janes TAT, Jessee JM, *et al.* Physiologically based pharmacokinetic (PBPK) modeling of metabolic pathways of bromochloromethane in rats. J Toxicol 2012; 2012629781
[http://dx.doi.org/10.1155/2012/629781] [PMID: 22719758]

[28]   Campbell J, Van Landingham C, Crowell S, *et al.* A preliminary regional PBPK model of lung metabolism for improving species dependent descriptions of 1,3-butadiene and its metabolites. Chem Biol Interact 2015; 238: 102-10.
[http://dx.doi.org/10.1016/j.cbi.2015.05.025] [PMID: 26079054]

CHAPTER 10

Green Chemistry: Making Chemistry Environment-Friendly

Sangeeta Bajpai[1,*], Saman Raza[2], Iqbal Azad[3] and Tahmeena Khan[3]

[1] *Amity School of Applied Sciences, Amity University, Lucknow, India*

[2] *Isabella Thoburn College, Lucknow, India*

[3] *Integral University, Lucknow, India*

Abstract: Chemistry is all around the universe. Green chemistry underpins the enormous social and technological changes in the future. Beginning from eco-friendly chemical synthesis to green catalyst *via* green chemical reactions, it finds a good correlation with the environment, taking biosynthesis and biomimetic principles into consideration. Widespread interest in this field is seen today among scientists. Considering the present scenario of "The age of tools", the compatibility with technology today is of utmost importance. Green chemistry is one of the powerful tools to cut the Gordian knot of pollution by reducing chemicals in the surroundings to make them eco-friendly. This chapter emphasizes the various aspects of green chemistry, from its principles to its applications, leading to a sustainable eco-friendly future.

Keywords: Chemistry, Environment, Green chemistry, Pharmaceuticals, Solvent, Sustainable.

INTRODUCTION

Green Chemistry

Keeping an eye on the past and looking at the present, the fact that chemistry is often misused, cannot be denied. The increase in the percentage of pollutants, toxic substances, and non-biodegradable materials all around, has resulted in imbalanced biodiversity. Applications of chemistry on living systems to increase productivity without damaging our natural resources are a very important aspect today. Chemical industries are of utmost importance in the world's economy; however, their success has defaced the environment to some extent. Fortunately, scientists are moving steadily towards attaining an eco-friendly approach by

* **Corresponding author Sangeeta Bajpai:** Department of Chemistry, Amity School of Applied Sciences, Amity University, Lucknow, India; E-mail: sbajpai1@amity.edu

Tahmeena Khan, Abdul Rahman Khan, Saman Raza, Iqbal Azad and Alfred J. Lawrence (Eds.)

practising resource sustainability. The chemistry community has been active for the last few years in developing various chemical strategies that have a less hazardous impact on biodiversity. This new approach of chemistry, called Green Chemistry, is also known by different names like Clean Chemistry, Environmentally Benign Chemistry, *etc*. The practice of chemistry in a manner that maximizes its benefits while reducing its adverse impacts has come to be known as green chemistry.

In a nutshell, green chemistry is an approach of promoting technologies or methodologies adopted for the generation/production of unhazardous chemical products essential for living systems, thereby removing pollutes from the environment. It is thus a potential driver of a sustainable ecosystem.

Green chemistry primarily involves the reduction of hazardous chemicals responsible for environmental damage. Green chemistry can also be defined as a sustainable process of minimizing hazardous substances, affecting biodiversity, from the environment economically. Today, green chemistry has generated some new terms *viz*. sustainable chemistry, atom economy, eco-efficiency, inherent safety, atom efficiency, ionic liquids, renewable energy resources, alternate feedstocks, *etc*.

Thus, today the need of the hour is developing various easy and economical methodologies for the production of harmless essentials required for mankind and the environment.

Principles of Green Chemistry

The twelve principles of green chemistry were introduced by Paul Anastas and John Warner in 1998. These principles provide guidelines for designing a chemical reaction to reduce the use of hazardous reagents and solvents and also avoid the formation of any toxic by-products. The principles are summarized below.

1. Prevention: It is well-known that many synthetic procedures involve the use of toxic reactants and solvents, and also produce a large number of toxic by-products. Many such reactions are carried out in industries. The chemical industry produces several million tonnes of such waste every year, which then requires clean-up. According to the green chemistry approach, it is better to prevent waste than to treat or clean up waste after it has been created. However, the absolute prevention of waste generation is virtually impossible in practice as the raw material cannot be fully utilized. Therefore, it is necessary to first consider whether the prevention of waste generation is possible and if not, methods should be devised to utilize the waste produced in the best possible way, so it becomes useful [1, 2].

2. Atom economy: Barry Trost introduced the concept of synthetic efficiency in 1990. Atom economy or atom efficiency refers to the concept of maximizing the use of raw materials [3, 4]. It is a theoretical value used to assess how efficient a reaction will be [5].

The ideal reaction would incorporate all the atoms of the reactants. Synthetic methods should be designed to maximize the assimilation of all materials used in the process into the final product. In this way, there is minimum waste formation. For example, cycloaddition reactions or multi-component coupling reactions constitute one category of efficient reactions (Fig. **1**).

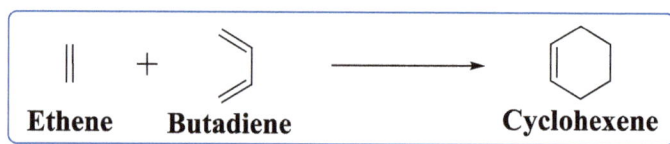

Ethene   **Butadiene**   **Cyclohexene**

Fig. (1). Diels Alder reaction showing 100% atom efficiency.

3. Less hazardous chemical synthesis: Synthetic procedures should be designed in such a way that less harmful reagents and solvents. Cascade or tandem reactions and enzymatic reactions are good examples of cleaner and more efficient synthetic pathways [5 - 13]. Harmful chemicals are now being replaced by cleaner and cheaper biological enzymes in many industrial processes [14]. For example, Dehydrogenases can be used for synthetic procedures involving dehydrogenation, like oxidation of alcohols. Enzymes like Oxidoreductases, Transferases, Hydrolases, Lyases, Dehydrogenases, and Isomerases, are now used in industries, such as pharmaceuticals, food, agrochemicals, cosmetics, *etc.* Also, the use of solvents and separation agents should be avoided wherever possible. For example, chromatographic separations use large quantities of solvents, which cause environmental pollution. Most traditional organic solvents are toxic, flammable, and corrosive, and their recycling is associated with considerable energy loss [1]. Therefore, green solvents like water should be used wherever possible.

4. Designing safer chemicals: One of the most challenging aspects of designing safer products is minimizing their toxicity while maintaining efficacy. For this, an understanding of not only chemistry but also toxicology and environmental science is required. Chemists often use highly reactive chemicals in synthesis procedures as they are useful in causing molecular transformations. However, they also react with unintended biological targets, resulting in unwanted adverse effects on humans and the environment. For this reason, the development of environment-friendly products, which would achieve their desired function while minimizing their toxicity, is required.

5. Safer solvents and auxiliaries: As mentioned above, solvents used in synthesis and separation may be toxic. Unnecessary use of these auxiliary substances should be avoided whenever possible and green solvents should be used where necessary. For example, benzene as a solvent must be avoided as it is carcinogenic. According to the principles of green chemistry, the choice of suitable alternatives for organic solvents is based on process safety, environmental safety, and sustainability of the process [1, 15 - 17]. Based on these parameters, scientists have now devised the use of solventless systems and green solvents like water, supercritical fluids, and ionic liquids [5, 18 - 20].

6. Design for energy efficiency: With increasing apprehensions over the depletion of petroleum resources and the growth in energy consumption, the development of energy-efficient synthetic procedures has become imperative [5, 21]. If possible, synthetic methods should be conducted at ambient temperature and pressure. Also, in a high-energy reaction, any unutilised energy may be considered a waste. Therefore, making changes like using methods of lowering the energy barrier of a chemical reaction (*e.g.* with the help of a suitable catalyst) or choosing suitable reactants so that the transformation may proceed at room temperature is desirable.

7. Use of renewable feedstock: Whenever technically and economically possible, renewable raw material or feedstock should be used. For example, biodegradable plastic materials and biodegradable packaging materials are now being developed and are in use. While bottles made of polyethylene (PE) blends are commonly used, bottles made of lactic acid polymer (PLA) that is made from lactic acid, produced by fermentation of dextrose obtained from starch, are also being produced and used by environment-friendly brands [22].

8. Reduce derivatives: Unnecessary derivatization, such as the use of blocking groups, protection/deprotection, and temporary modification of physical/chemical processes, should be avoided if possible or minimized if essential. This is because such steps require additional reagents and lead to the generation of waste. Here again, enzymes can be used because being specific in their action, they can react with one group only, and hence protecting groups may not be required. An example of the use of enzymes to avoid protecting groups is the industrial synthesis of semi-synthetic antibiotics, such as ampicillin and amoxicillin. In the industrial synthesis, Penicillin G (R=H) is first protected as its silyl ester [R =

Si(Me)$_3$] and then reacted with phosphorus pentachloride to form the chlorimidate **1**. The hydrolysis of 1 gives the desired 6-APA from which semi-synthetic penicillins are manufactured(Fig. **2**).

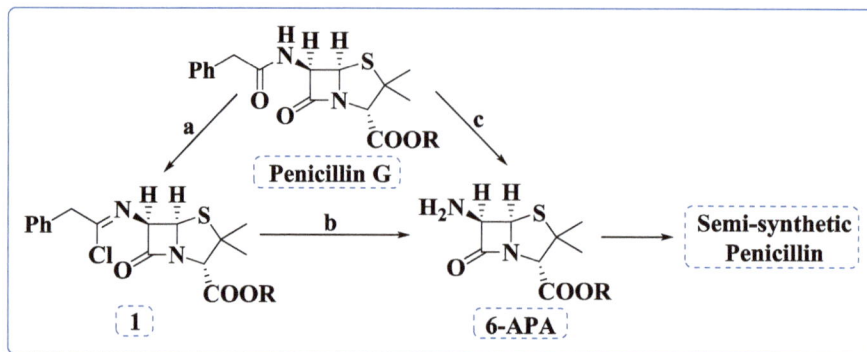

Fig. (2). Industrial synthesis of penicillin (i) TMSCl then PCl$_5$, PhNMe$_2$, CH$_2$Cl$_2$, -40°C (ii) n-BuOH, -40°C, then H$_2$O, 0 °C (iii) Pen-acylase, water.

This synthesis has now been replaced by an enzymatic process using pen-acylase. This synthesis is carried out in the water. This is a green alternative to the earlier method as chemicals like the silyl protecting group are not required [23].

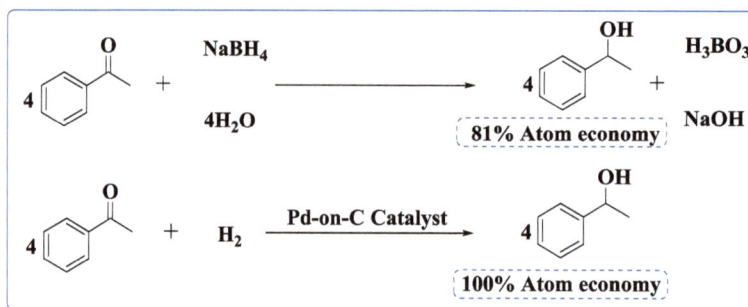

Fig. (3). Example of increasing Atom Economy with a catalyst.

9. Catalysis: Catalytic reagents are superior to stoichiometric reagents as the use of a traditional stoichiometric amount of reagents is linked to waste formation [1, 5]. Catalysts help in increasing the atom economy of the reaction. For example, in the reduction of a ketone to the corresponding secondary alcohol, when the reaction is carried out using sodium borohydride the atom economy is 81%. However, when the same reaction is carried out using Platinum as the catalyst and molecular hydrogen as the reductant, the atom economy is 100% as there is no waste (Fig. **3**) [23]. Also, biocatalysts *i.e.* enzymes are highly useful in green synthesis.

10. Design for degradation: Chemical products should be designed in such a way that they break down into harmless degradation products and do not persist in the environment. Chemicals can be classified as PBT (Persistent, Bioaccumulative, Toxic), based on their degradation level. Chemical products can be made less persistent by making changes in the structure to enable biodegradation, hydrolysis, and photolysis. For example, in the 1950s, tetrapropylene-alkylbenzene sulfonate (TPPS) was used as a surfactant in laundry detergents. Due to its incomplete degradation, it would accumulate in the water supply and the water tended to foam when coming out of the tap. The issue was resolved by replacing the methyl branched chain of TPPS with a linear carbon chain that reduced the bio-persistence of the chemical. We know that some chemical structures such as branched chains, halogens, quaternary carbons, and certain heterocycles usually have enhanced persistence while groups like esters or amides, which are recognised by enzymes present in microbes, may help in the degradation of the molecules. Therefore, extensive knowledge about the link between structural features and the degradability of molecules is needed at the time of synthesis to predict their ease of degradation. This can be achieved with the help of databases of molecules, models that evaluate biodegradability or PBT attributes, and experimental testing [23].

11. Real-time analysis for pollution prevention: Analytical methodologies need to be further developed to allow for real-time, in-process monitoring and control, before the formation of hazardous substances. Green analytical chemistry involves the use of analytical procedures that produce lesser waste and are safer for the environment and human health [5, 24]. The products used in the manufacture of analytical apparatus should also be taken into consideration. For example, mercury electrodes are often used in electrochemistry. The use of toxic mercury needs to be reduced. This can be done by replacing mercury electrodes with carbon-based electrodes such as nanotubes or nanofibers [5, 25]. The effective application of process analytical chemistry directly contributes to the safe and efficient operation of chemical plants worldwide.

12. Inherently safer chemistry for accident prevention: Substances used in a chemical process should be chosen to minimize the potential of chemical accidents, such as chemical releases, explosions, and fires. According to the chemical accident prevention and the clean air act amendments of 1990, avoiding accidents starts with identifying the hazards [5, 26]. All kinds of hazards whether it is toxicity or physical hazards, such as flammability, should be considered. The Bhopal gas tragedy is a reminder to the scientific community to replace hazardous chemicals and processes with safer options. For example, the increasing use of supercritical CO_2 in place of organic solvents is non-toxic, non-explosive, and environment-friendly.

Applications of Green Chemistry

1. Corrosion

The term corrosion defines the deterioration of a material due to its reaction with the environment or incompatible chemicals. Metallic deterioration grows very fast due to destructed passive barrier which alters the behaviour of both the constituents and the environment. Formation of an oxide layer, changes in electrochemical potential, cation insertion in coating matrix, *etc.* are some of the examples occurring due to the passive barrier destruction. The study of metallic corrosion has become a subject of intangible and practical concern as to achieve envisioned purpose, by inducing acid cleaning, marinating, and de-scaling processes on the metallic surfaces with the use of corrosion inhibitors.

Metallic alloys like aluminium alloys, brass, copper and mild steel are the eminent chemicals actively and commonly used for major industrial operations, such as in carriages, distribution, storage containers *etc.* Various theories envisaging corrosion protection are known. The most acceptable theory was proposed by Whitney in 1903 [27]. The other theories are acid theory [28], colloidal theory [29] and direct chemical attack theory [30].

In industries, corrosion of iron alloys is an area of great focus, due to their good compatibility, low cost and good mechanical properties. Recently, studies have come forward, employing a green approach by using green inhibitors for attaining control measures for corrosion-induced wastages. Leaves of *Cardiospermum helcabum* (CH), *Gymenema sylvestre* (GS) and *Wrightia tinctorial* (WT) have been used as green inhibitors for corrosion control (Fig. **4**) [31].

Fig. (4). Green inhibitors-Cardiospermum helcabum plant (a), Gymenema sylvestre (b), Wrightia tinctorial (c) [31].

The phytochemical preliminary analysis of CH showed the presence of six compounds *i.e.* cardiac glycosides, flavonoids, alkaloids, saponins, steroids and tannins. Flavonoids proved to be the most active ones (Fig. **5a**). In WT two active compounds, wrightial and flavonoid were found with prominent activity (Fig. **5b**).

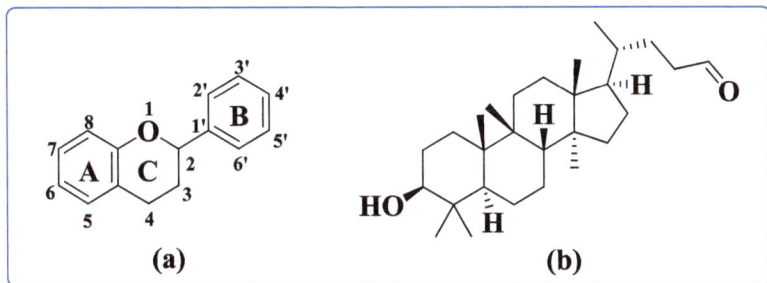

Fig. (5). Chemical structures of phytochemicals present in CH and WT.

The green inhibitors were extracted from CH, GS and WT plants by reacting with alcohol. 0.1 N HCl (prepared by using double distilled water) was used as the electrolyte. Experimental studies proved the active role of these green inhibitors. In weight loss studies [using the formula Corrosion rate {mmpy} = 87.6 x W/ DAT {where mmpy = millimetre per year, W= mass loss (mg), D = Density (gm/cm3), A = Area of specimen (cm2) }], the findings revealed a moderate decrease in corrosion rate with increase in concentration of CH inhibitors, and 91.67% efficiency was achieved at 1000 ppm of these inhibitors. Temperature studies revealed a decrease in corrosion rate from 15.2083 to 2.7157 mmpy, at 313 K, with maximum efficiency being 82.14% [31]. Similar results were obtained with GS and WT leaves-extract as green corrosive inhibitors.

2. Nanotechnology

Clean biodiversity is the need of the hour today and the green chemistry approach is the solution. Most of the upcoming researches are inculcating the green approach in their methodology. The nanotechnology field is also not untouched. Green nanomedicines have emerged in drug delivery systems using polymers, biomaterials (plant extracts, microorganisms), *etc*. Referring to the twelve principles of green chemistry, only a few research articles are available on nano-drug delivery systems, considering these principles. Nanometal drugs find great potential in drug delivery systems. Maghaemite and Magnetite nanoparticles are well-known nanometal drugs used in the imaging and treatment of tumours, and therapeutic purposes [32 - 34]. A modified methodology has been adopted recently, for a nano iron oxide drug, found effective in gene therapy, chemotherapy, *etc*. by minimizing the destruction in the blood-brain barrier, thereby escalating the safe and effective delivery of nano iron oxide to the brain

[35]. Table **1** is a list of nano-metals in the drug delivery process inculcating green chemistry principles.

Green and potent plant-mediated synthesis of gold nanoparticles, using Indian Banyan leaf extract [37], and silver nanoparticles, using aqueous extract of *Garcinia mberti* (an endangered species) [38], is reported. These nanoparticles show antibacterial properties and are listed as good therapeutic agents against human microbes.

Table 1. Nano-metal drugs obeying Green Chemistry Principle [36]

Nano metals in drug delivery	Use	Green Chemistry Principles
Bio-corrodible iron stents	Porcine coronary arteries	Designing safer chemicals
Basal fibroblast growth factor-iron oxide nanoparticles	Cancer radiation therapy	Designing safer chemicals
Magnesium nanoparticles	Hyperthermia therapy	Design for degradation (biodegradable)
Iron oxide coating with Polyethylene glycol	Targeting agent	Atom economy
Hollow mesoporous silica nanoparticle--cyclodextrin nanocarrier	Inhibited tumour growth	Less hazardous chemical syntheses
Silver nanoparticles in a polymer	Antiseptics and antimicrobial	Safer solvents and auxiliaries
Molybdenum disulfide Nanosheet	Photothermal agent	Less hazardous chemical synthesis

3. Pharmaceuticals

The green approach in pharmaceuticals has proven to be a revolutionary contribution to the decrease in suffering and death. The key focus of pharmacists today is the development of various creative measures for minimizing side effects. The pharmaceutical industry is at the forefront for inducing modifications, with alternative and innovative ideas for producing greener feedstocks and solvents, thereby stepping ahead towards sustainability.

The overall toxicity profile of a drug is governed by its mass and solvents. Pharmaceutical industries are known for the use of large amounts of solvents, although a chemist, in a retrosynthetic analysis, has an interest in the reactivity and building up of the product molecule rather than the use of the solvent. Batch chemical operations, found typically in the pharmaceutical industry, show the excess need for solvent [39].

Zhang *et al*. reported [40] a review on green solvents used in the pharmaceutical industry as well as green medicines. The selection of chemicals in the synthesis of pharmaceuticals (a wide area that includes medicinal drugs, disinfectants, dietary supplements, diagnostic, packaging materials, *etc*.) is being extensively researched day by day [41]. Some selective case studies related to greener technological approach/ pathways applied in the fields of solvents and catalysts, and also drug synthesis resulting in a reduction of toxic by-products, thereby increasing the yield of the products, are discussed here to showcase some applications of green chemistry in the pharmaceutical industry [42].

Efficient and economic routes applied for Oligonucleotides Drugs [43]

Modifications in conventional theories of medicine have emerged with safe and effective therapeutic ones. The work done in this area obeys the maximum laws of green chemistry. The protocols formulated, developed, or modified are well acquainted with green chemistry. Let us take the example of Phosphoramidite (a blend of antisense oligonucleotides) which has upgraded the atom economy and cost-effectiveness by discarding toxic materials, following the seventh principle of green chemistry, which states that formulations be developed for renewing raw feedstocks without depleting. In the process of synthesis of oligonucleotides drugs, the first step involves the replacement of fish-derived nucleosides with synthetic nucleosides. The first-generation antisense drugs 2'-deoxyoligonucleotides have uniform phosphate diester backbones (Fig. **6**). The key building blocks for the assembly of 2'-deoxyoligonucleotides are 2'-deoxynucleosides, historically obtained from DNA salt *via* enzymatic action from the milt of marine sources. Due to the poor availability of the source and expensive isolation process chemists have developed a new methodology for the production of efficient 2'-deoxygeneration protocols for the conversion of readily available ribonucleosides to 2' deoxynucleosides, which are fully independent of marine DNA resources.

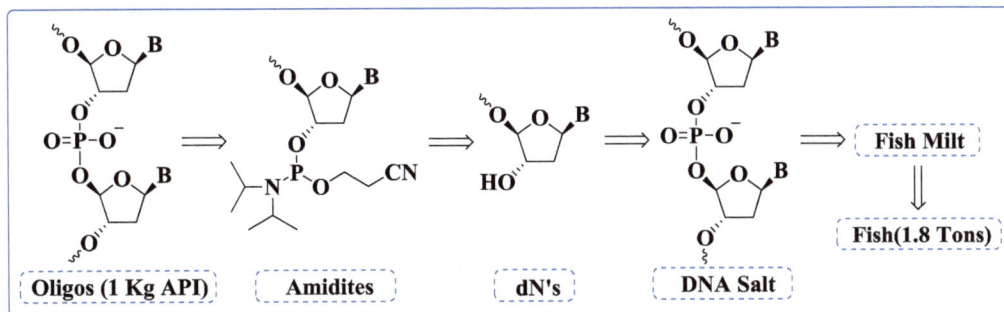

Fig. (6). Raw material generation for oligonucleotides [43].

The Antisense oligonucleotides involve Phosphoramidite coupling chemistry [44]. The yield and quality of oligonucleotides adopting this protocol depend on four key synthetic steps (Fig. **7**) Each step involves the critical elimination of acid-labile 5'-O-DMT-(4,4'-dimethoxytrityl) protecting group with halo-acetic acid, using dichloromethane as solvent. Being toxic, carcinogenic, and volatile, dichloromethane is less frequently used. Toluene is used as a green substitute for dichloromethane [45]. Toluene is known to be a very "widely accepted industrial solvent" due to its lower vapour pressure and eco-friendly behaviour [46].

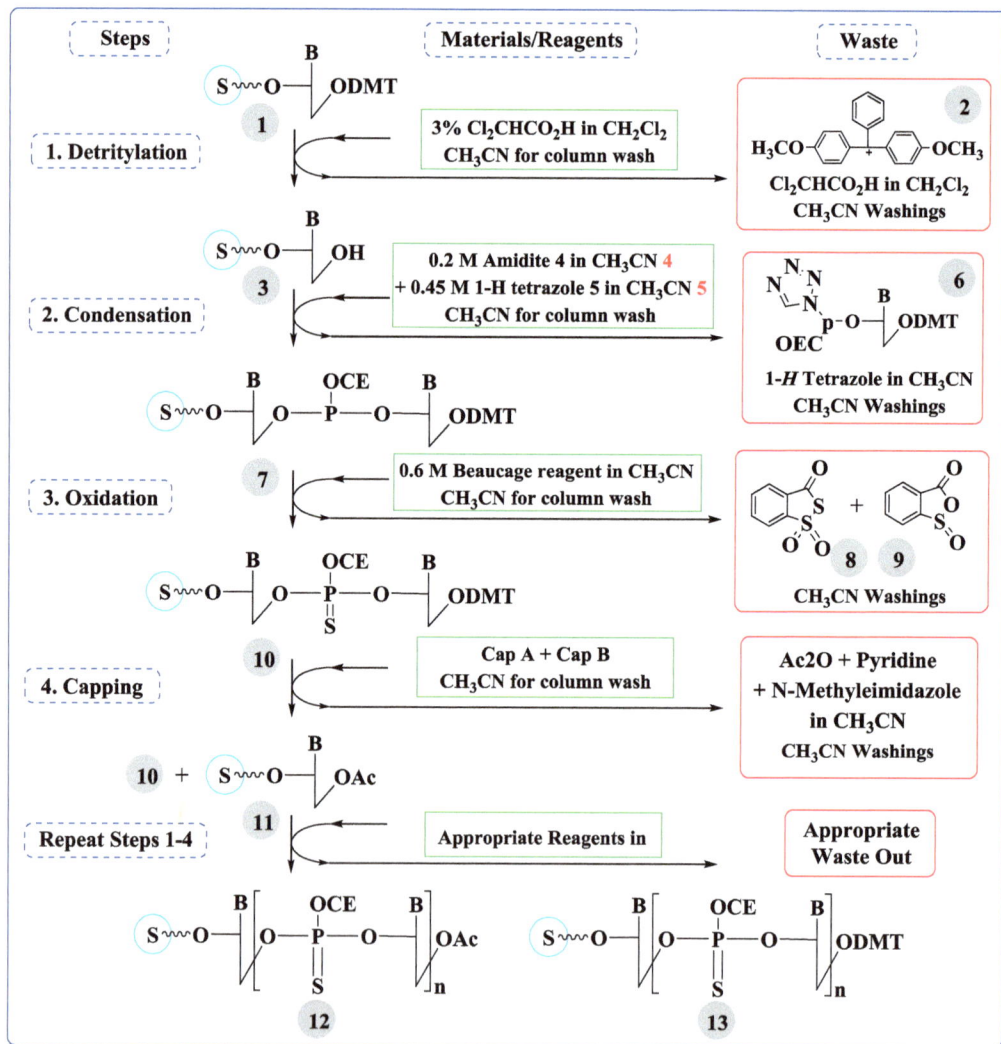

Fig. (7). Schematic representation of key Oligonucleotides synthesis, using standard reagents [43].

Analgesic and Anti-inflammatory Drugs

Paracetamol (Acetaminophen) is a well-known drug used as a pain reliever and for controlling flu and cold. Its synthesis involves a three steps reaction using phenol. It can be synthesized from phenol in three steps. The first step involves electrophilic aromatic substitution on phenol with nitric acid to create p-nitrophenol, which on hydrogenation using iron as a catalyst, produces p-aminophenol and finally, paracetamol is synthesized by acylation of the aminophenol.

Paracetamol from benzene (Scheme **1**). It involves four routes

-Route 1: Nitration of chlorobenzene.
-Route 2: Nitration of phenol.
-Route 3: Reduction of nitrobenzene.
-Route 4: Acetamidation of hydroquinone.

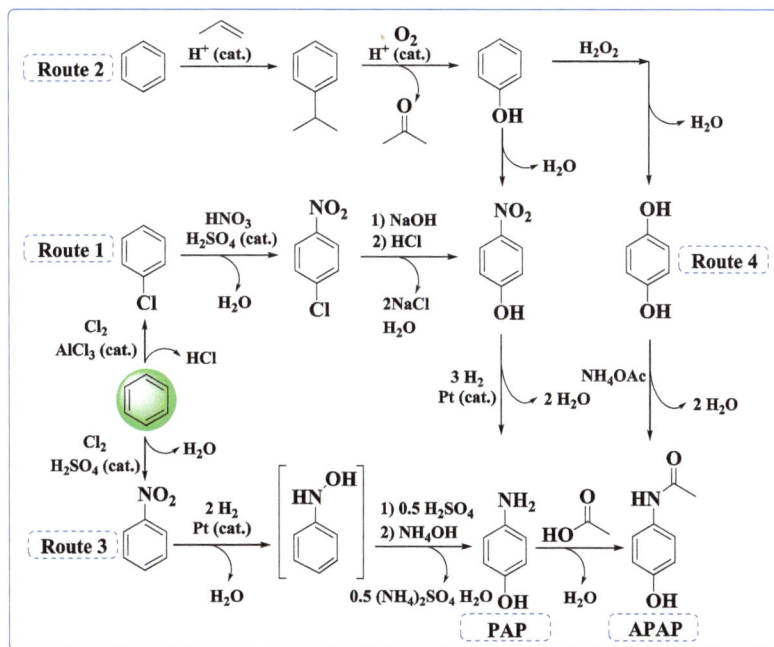

Scheme 1. Different routes for the synthesis of Paracetamol [47].

The theoretical Atom economy (AE) and E factors are reported following Trost's - [3, 4] and Sheldon's [48] definitions of AE and E respectively. The results of theoretical atom-economies and E factors are compiled in Table **2**.

The figures in Table **2** predict that route 1 (nitration of chlorobenzene) is the least attractive in terms of atom economy and E factor as compared to routes 2, 3 and

4. Moreover, these values are not enough for selecting the above procedure as the 'true' best method. Several other parameters also have to be considered. This new methodology, including the green step, minimized chemical waste [47]. Synthesis of Ibuprofen, a commonly used anti-inflammatory drug, is another example under this category [49].

Table 2. AE and E factor of considered routes for paracetamol synthesis from benzene [47].

Route	Considered waste (equiv.)	Waste mol. wt. (g/mol)	AE (%)	E factor[*]
1	H_2O (5)	18.02 (x5)	38	1.61
	NaCl (2)	58.44 (x2)		
	HCl (1)	36.46		
2	H_2O (4)	18.02 (x4)	54	0.86
	Acetone (1)	58.08		
3	H_2O (4)	18.02 (x4)	52	0.91
	$(NH_4)_2SO_4$ (0.5)	132.14 (x0.5)		
4	H_2O (3)	18.02 (x3)	57	0.74
	Acetone (1)	58.08		

*For theoretical E factor calculations, water was considered as waste.

Another example is Quinapril (Quinapril hydrochloride), which acts as an angiotensin-converting enzyme inhibitor and is used for curing hypertension [50]. The reactants used in its synthesis are methylene chloride, a potentially violent hydroxy-benzotriazole, dicyclohexyl-carbodiimide [DCC], and a large amount of toluene is used for separating the acetic acid formed, by the solvent-exchange process. The greener route for this synthesis begins with changing the starting materials to N-carboxyanhydride and isoquinoline carboxylic acid t-butyl ester salt, which react by direct amide coupling. In this way, DCC, its waste product dicyclohexylurea (DCU), and the need for a chlorinated solvent are avoided. The ester is hydrolysed by a minimum amount of CH_3COOH/HCl, the salt formed is isolated by adding the antisolvent acetone. In this way, a lengthy solvent exchange procedure is avoided and the overall yield increases from 58% to 90%, with minimum waste production and use of greener solvents.

Use of Solvents

Solvents are usually liquid at room temperature. They can dissolve substances and other materials. They are generally used in cleaning and separation processes. Solvents are mainly classified as aqueous and organic. They are invaluable substances for pharmaceutical chemists as well as in industrial practices. Different functional groups attached to the solvent leads to a change in their characteristics. They are generally used at the purification stage or in the formulation. Solvents are used in great quantity; that is why the environmental concern associated with them is more. Approximately half of the material used in the pharmaceutical production units is solvent [51]. Despite all the benefits, solvents do not directly influence the end product of the reaction and remain unconsumed, hence the use of these solvents, many of which are toxic to the environmental health, should be checked or a better alternative must be used which does not cause unnecessary pollution and damages health.

Solvent Pollution and Hazardous Environmental and Health Effects

Many a time, the otherwise hazardous properties of solvents are their good qualities. The volatility of a solvent during distillation and purification is appreciated, even though the same virtue can cause serious air pollution and risks to the exposed. Solvents with high polarity can lead to reproductive malfunctions [51]. Hydrocarbon-based solvents provide excellent solubility to oils, but they are highly flammable and owing to their poor water solubility, may lead to bioaccumulation and toxicity to aquatic flora and fauna. Solvents, such as benzene, can cause serious effects as they interact with crucial biochemical processes in the human body [52] leading to adverse effects on respiratory, endocrine, and haematopoietic systems.

The haematopoietic system is highly sensitive to pollutants and toxic substances and the exposure of toxic solvents may hinder the process of red blood cell proliferation, leading to impaired heme synthesis and life expectancy of RBCs, in some cases causing aplastic anaemia [53].

The thyroid gland is the main target of the endocrine system to be affected by solvents. The hormones of the hypothalamic-hypophysial axis are controlled by several factors and are sensitive to toxic substances, including solvents [54]. It has been found that workers exposed to benzene may have alterations in the level of TSH and T3, T4, and T4H. Solvents such as methylene chloride, which is used in industrial processes, are a known carcinogen. Trichloroethylene, which is used as a solvent in groundwater, can damage the liver, lungs, and nervous system.

Other than these, chloroform, and carbon tetrachloride cause liver toxicity. Glycol ethers and other chlorinated solvents cause kidney damage. Chronic exposure to carbon disulphide causes coronary heart disease. Some solvents lead to complications in cardiovascular functioning, which arise due to the increased sensitivity of the muscle of the heart. Some organic solvents may cause a placental barrier and may be teratogenic, like glycol, ethers, and ethanol.

Once solvents are released in the environment, they may partition into water, air and soil phases. Solvents may also be transformed into other compounds which are less or more problematic.

Toluene contributes to the formation of tropospheric ozone at an urban scale. Other solvents similarly can have a plethora of impacts, ranging from local contamination to variations of the global climate. The difference in the impact caused by the solvent is based on its chemical properties and reactions that the solvent may undergo in the atmosphere, how it is released, and the variations in spatial and temporal scales of atmospheric phenomena. From an environmental perspective, solvents like benzene, xylenes, formaldehyde, dichlorobenzene, and oxygenated organic solvents including ethers, alcohols, and ketones are important. Some solvents also lead to the formation of secondary pollutants like tropospheric ozone and particulate matter (PM), which cause serious health damage and environmental deterioration on a larger scale. Slowly reacting solvents cause imbalances in living systems.

Green Solvents as Environment-friendly Alternatives

Owing to the associated challenges with the use and release of solvents, pharmaceutical, and industrial practices are promoting green chemistry. US Environmental Protection Agency (EPA) has recommended green chemistry as an innovative strategy to reduce toxic and undesired waste and associated health impacts. Not only pharmaceutical companies, other chemical industries as well are taking advantage of green chemistry practices for decreasing the waste amount and cost and for environmental and health-related benefits. It is also estimated that green chemistry can save the industry approximately USD 65.5 billion by 2020 by the reduction of manufacturing costs [55].

Solvents are very important when it comes to the implementation of greener strategies and are under consideration by several pharmaceutical companies, such as Sanofi-Aventis [56] and GlaxoSmithKline [57]. These companies have promoted the conversion of hazardous solvents such as halogenated solvents and petroleum-based to be replaced with greener solvents, such as glycerol and ethyl lactate [58]. Green solvents include water [59], supercritical fluids [60], gas expanded liquids [61], ionic liquids [62], liquid polymers [63] and solvents derived from biomass [64]. Many solvent selection guides have been formulated by different pharmaceutical industries like GSK, Pfizer, and Sanofi. The first of these guidelines were given by SmithKline Beecham [65]. The guidelines were based on the impact on incineration, like the heat of combustion, emission, water-solubility, boiling points, vapour pressure, *etc*. Thirty-five solvents were ranked according to these criteria. Afterward, regulatory concerns were added to the guide published by GSK [66]. The Pfizer solvent selection guide categorized solvents as preferred, usable and undesirable [67]. The guidelines were based on workers' safety and health concerns related to the solvent like carcinogenicity, reproductive toxicity, skin sensitivity, and flammability of the solvent, *etc.* The guide developed by Sanofi categorised solvents in different classes like alcohols, ketones, esters, ethers, hydrocarbons, halogenated, polar aprotic, *etc* [68]. The basis of the ranking was safety, occupational health, environmental health, quality, and industrial constraints. Some examples of the green solvents include-

Water

Water replaces many toxic and hazardous substances and is also used as a solvent in many organic reactions like in the synthesis of benzothiazoles, benzothiazoline, chromenoisoxazole, *etc* [69]. Important reactions like Wittig, which is pertinent to pharmaceutical chemistry, have been used in the formation of a new C-C bond. A green protocol for Wittig reaction in an aqueous medium has been carried out at 25 °C [70]. 2,2'-dipyrromethane has been synthesized using pyrrole and diethyl ketone in water [71]. The product was obtained in a yield of up to 80%. Isocoumarin has also been synthesized in H_2O by the reaction of salicylic acid and alkyne [72].

For the preparation of lactam, water has been as a solvent in which acetoacetic acid and isonitrile react with an amine. The reaction took two hours to complete and the product was obtained in 93% yield [73]. Besides lactam, lactone is another crucial ring that is present in many natural products and used in many drugs. The yield of the product was more when cyclization of allyl-iodoacetate was carried out in triethyl borane in water [74].

The conversion of CH_2 group of isatin into oxime group has also been achieved in water. Isatin is a building block of several bioactive moieties [75].

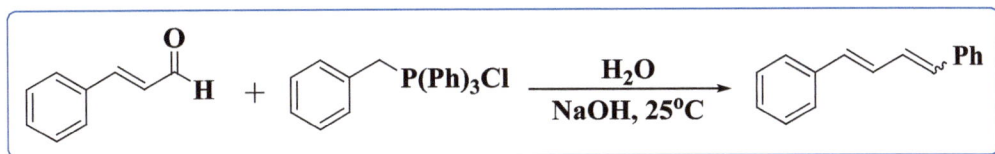

Scheme 2. Wittig reaction in water.

Glycerol

Glycerol is another non-toxic solvent that has high polarity and is available at a low price. It is also used in the reduction process of different carbonyl compounds [76]. Glycerol is a polyalcohol and finds use in many pharmaceutical and food industries. Safae *et al.* have synthesized 4H-pyrans using a one-pot strategy in a catalyst-free medium, using glycerol as a medium [77]. Medicinal scientists have used cyclization reaction using glycerol as a drug-design strategy. Medicinal scientists have carried out the nucleophilic attack on the beta position of α,β-unsaturated carbonyl molecules in glycerol. This reaction is useful in the derivatization of lead compounds in pharmaceutical industries. Using crude glycerol increased the yield up to 81% (Scheme **3**) [78]. Alkylated derivatives of glycerol have also been prepared and evaluated for further applications [79]. Jerome *et al.* have reported some common synthetic reactions on glycerol and they have discussed yield, reaction time, and sustainability of glycerol.

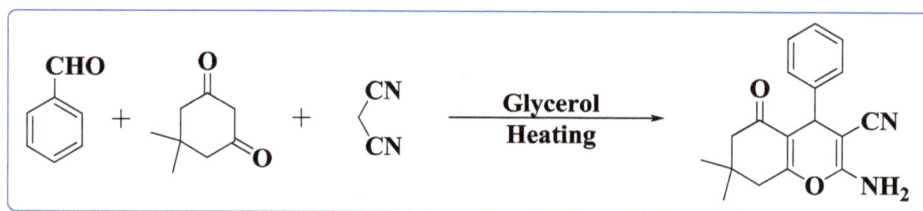

Scheme 3. One-pot three-component reaction in glycerol.

Supercritical Carbon Dioxide

Owing to its low toxicity and environment-friendly impact, supercritical carbon dioxide $ScCO_2$ has gained popularity. It represents a fluid state of CO_2 where it is held at or above its critical temperature (304.25 K, 31.10 °C, 87.98 F) and critical pressure (72.9 atm, 7,39 mPa, 1,071 psi) [80]. Reactions like hydrogenation, epoxidation, radical reactions, palladium mediated C-C bond formation, ring closure, polymerisation, and many other reactions can be carried out in $ScCO_2$ [81].

Carbonic Esters

Carbonic esters like DMC, dimethyl carbonate ($CH_3OCOOCH_3$) are also considered green solvents. They replace methyl chlorides and dimethyl sulphate esters which are toxic and hazardous. DMC can also be used in the methylation of phenols. DBU is another solvent that can be used for methylation reactions of phenols, indoles, and benzimidazoles.

Vegetable Oils

Vegetable oils are extracted from the seeds of plants. Chemically they are triglycerides. They are used in the synthesis of biopolymers. Acyclization and cyclization reactions have been carried out in corn oil (Scheme **4**). This is the first example of a reaction using vegetable oil as the medium [82].

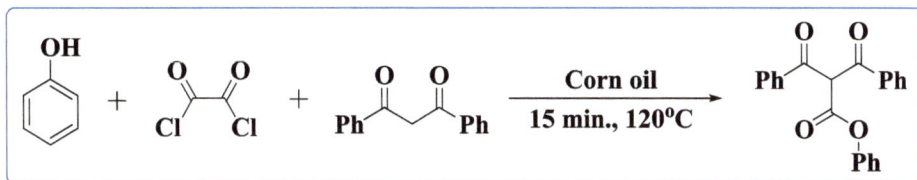

Scheme 4. Acyclization in corn oil.

Ionic Liquids

Ionic liquids are organic salts that are liquid at ambient temperature. They are non-volatile, non-flammable, and thermally and chemically stable, making them better alternatives in green chemistry. Owing to their high polarity, they are used in many biochemical reactions. The most interesting part of ionic liquids is the alterations that can be made in the attached cations, anions, and alkyl groups (Fig. **8**) [83]. Ionic liquids are more viscous and to adjust the physical properties, some substitutions are made with hydrogen donors, such as glycerol, oxalic acid, and urea. Ionic liquids are also called deep eutectic solvents (DES) (Fig. **9**) when they are attached with quaternary ammonium salts and a hydrogen bond donor [84].

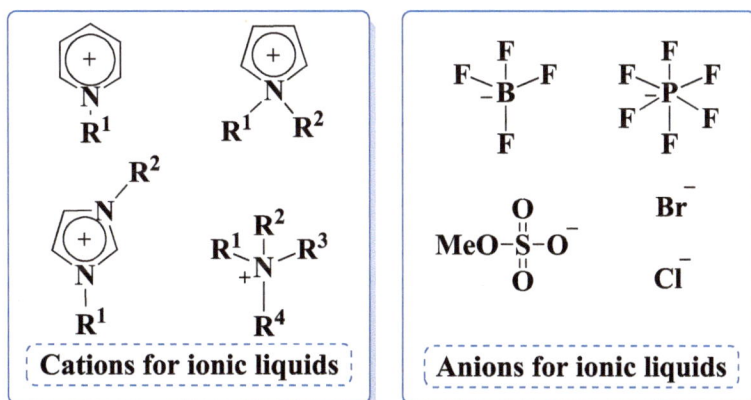

Fig. (8). Some common cations and anions for ionic liquids.

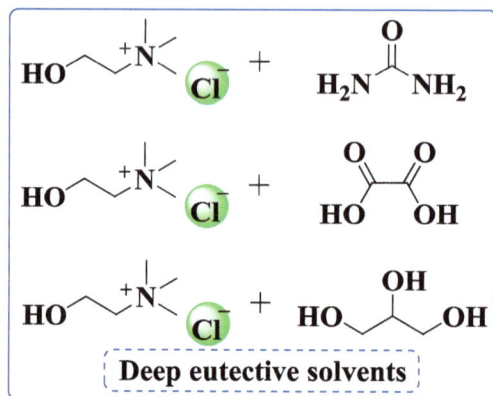

Fig. (9). Some examples of deep eutectic solvents.

CONCLUDING REMARKS

Thus, chemistry no doubt has invented many beneficial products for human wellbeing, but the formation of harmful and undesirable waste products cannot be ignored. Green chemistry provides a big platform to avoid these undesirable contaminating substances. It opens a multifaceted and wide research scope for the invention of more efficient processes to minimize waste and maximize the yield of the desired product. Although green chemistry alone cannot reduce these impacts, it is certain that by employing the twelve principles of green chemistry, as given by T. Poul, Anastas, and J.C. Warner, sustainable ecological balance can be achieved. Enormous efforts are being made by keeping this in mind to design new and superlative strategies, resulting in a greener world.

CONSENT FOR PUBLICATION

Not Applicable.

CONFLICT OF INTEREST

The author confirms that this chapter contents have no conflict of interest.

ACKNOWLEDGEMENT

Declared none.

REFERENCES

[1]   Ivanković A, Dronjić A, Bevanda AM, Talić S. Review of 12 Principles of Green Chemistry in Practice. Int J Sustain Green Energy 2017; 6: 39-48.
[http://dx.doi.org/10.11648/j.ijrse.20170603.12]

[2] Valavanidis A, Vlachogianni T, Fiotakis K.

[3] Trost BM. The atom economy--a search for synthetic efficiency. Science 1991; 254(5037): 1471-7.
[http://dx.doi.org/10.1126/science.1962206] [PMID: 1962206]

[4] Trost BM. Atom Economy- A Challenge for Organic Synthesis: Homogeneous Catalysis Leads the Way Angew. Chem Int Edit English 1995; 34: 259-81.
[http://dx.doi.org/10.1002/anie.199502591]

[5] Anastas P, Eghbali N. Green chemistry: principles and practice. Chem Soc Rev 2010; 39(1): 301-12.
[http://dx.doi.org/10.1039/B918763B] [PMID: 20023854]

[6] Jørgensen KA, Ed. Cycloaddition reactions in organic synthesis. John Wiley & Sons 2002.

[7] Flick AC, Padwa A. A conjugate addition-diploar cycloaddition approach towards the synthesis of various alkaloids. ARKIVOC 2011; vi: 137-61.

[8] Smith MB, March J. March's Advanced Organic Chemistry: Reactions, Mechanisms, and Structure. John Wiley & Sons 2019.

[9] Knipe AC, Watts WE, Eds. Organic Reaction Mechanisms. John Wiley & Sons 1997.
[http://dx.doi.org/10.1002/9780470066935]

[10] Zhu J, Bienaymé H, Eds. Multicomponent reactions. John Wiley & Sons 2006.

[11] Touré BB, Hall DG. Natural product synthesis using multicomponent reaction strategies. Chem Rev 2009; 109(9): 4439-86.
[http://dx.doi.org/10.1021/cr800296p] [PMID: 19480390]

[12] Jacobi von Wangelin A, Neumann H, Gördes D, *et al.* Unusual coupling reactions of aldehydes and alkynes: a novel preparation of substituted phthalic acid derivatives by automated synthesis. Chemistry 2003; 9(10): 2273-81.
[http://dx.doi.org/10.1002/chem.200204668] [PMID: 12772302]

[13] Nicolaou KC, Montagnon T, Snyder SA. Tandem reactions, cascade sequences, and biomimetic strategies in total synthesis. Chem Commun (Camb) 2003; 2003(5): 551-64.
[http://dx.doi.org/10.1039/b209440c] [PMID: 12669826]

[14] Sheldon RA. Metrices of Green Chemistry and Sustainability: Past, Present, and Future. ACS Sustain Chem& Eng 2017; 6: 32-48.
[http://dx.doi.org/10.1021/acssuschemeng.7b03505]

[15] Hoffert MI, Caldeira K, Benford G, *et al.* Advanced technology paths to global climate stability: energy for a greenhouse planet. Science 2002; 298(5595): 981-7.
[http://dx.doi.org/10.1126/science.1072357] [PMID: 12411695]

[16] Murariu M, Dubois P. PLA composites: From production to properties. Adv Drug Deliv Rev 2016; 107: 17-46.
[http://dx.doi.org/10.1016/j.addr.2016.04.003] [PMID: 27085468]

[17] Draths KM, Frost JW. Green Chemistry: Frontiers in Benign Chemical Syntheses and Processes. New York: Oxford University Press 1998; p. 150.

[18] Constable DJ, Curzons AD, dos Santos LMF, *et al.* Green chemistry measures for process research and development. Green Chem 2001; 3: 7-9.
[http://dx.doi.org/10.1039/b007875l]

[19] Constable DJ, Curzons AD, Cunningham VL. Metrics to 'green' chemistry—which are the best? Green Chem 2002; 4: 521-7.
[http://dx.doi.org/10.1039/B206169B]

[20] Kerton FM, Marriott R. Alternative solvents for green chemistry (No 20). Royal Society of Chemistry 2013.

[21]   Laughton MA. Renewable Energy Sources: Watt Committee: report number 22. CRC Press 1990.

[22]   Madhavan Nampoothiri K, Nair NR, John RP. An overview of the recent developments in polylactide (PLA) research. Bioresour Technol 2010; 101(22): 8493-501.
[http://dx.doi.org/10.1016/j.biortech.2010.05.092] [PMID: 20630747]

[23]   https://www.acs.org/content/acs/en/greenchemistry/principles/12-principles-of-green chemistry.html

[24]   Keith LH, Gron LU, Young JL. Green analytical methodologies. Chem Rev 2007; 107(6): 2695-708.
[http://dx.doi.org/10.1021/cr068359e] [PMID: 17521200]

[25]   Mohammad WA, Ali A. Q.; Errayes, A.O. Green Chemistry: Princples, Applications, and Disadvantages. Chem Methodologies 2020; 4: 408-23.
[http://dx.doi.org/10.33945/SAMI/CHEMM.2020.4.4]

[26]   Stellman JM, Ed. Encyclopaedia of occupational health and safety. International Labour Organization 1998; p. 3.

[27]   Whitney WR. The corrosion of iron. J Am Chem 1903; 25(4): 394-406.
[http://dx.doi.org/10.1021/ja02006a008]

[28]   Walkar WH, Cederholm AM, Bent LN. The corrosion of iron and steel. J Am Chem Soc 1907; 29: 1251-64.
[http://dx.doi.org/10.1021/ja01963a001]

[29]   Garner M, Reglinski J, Smith WE, Stewart MJ. The interaction of colloidal metals with erythrocytes. J Inorg Biochem 1994; 56(4): 283-90.
[http://dx.doi.org/10.1016/0162-0134(94)85108-5] [PMID: 7844588]

[30]   Barik RC, Wharton JA, Wood RJK, *et al.* Erosion and erosion–corrosion performance of cast and thermally sprayed nickel–aluminium bronze. Wear 2005; 259: 230-42.
[http://dx.doi.org/10.1016/j.wear.2005.02.033]

[31]   Deivanayagam P. 2019.

[32]   White MA, Johnson JA, Koberstein JT, Turro NJ. Toward the syntheses of universal ligands for metal oxide surfaces: controlling surface functionality through click chemistry. J Am Chem Soc 2006; 128(35): 11356-7.
[http://dx.doi.org/10.1021/ja064041s] [PMID: 16939250]

[33]   Arruebo M, Fernandez PR, Ibarra MR, *et al.* Magnetic nanoparticles for drug delivery. Nano Today 2007; 2: 22-32.
[http://dx.doi.org/10.1016/S1748-0132(07)70084-1]

[34]   Mornet S, Vasseur S, Grasset F, Duguet E. Magnetic nanoparticle design for medical diagnosis and therapy. J Mater Chem 2004; 14: 2161-75.
[http://dx.doi.org/10.1039/b402025a]

[35]   Sun Z, Worden M, Thliveris JA, *et al.* Biodistribution of negatively charged iron oxide nanoparticles (IONPs) in mice and enhanced brain delivery using lysophosphatidic acid (LPA). Nanomedicine (Lond) 2016; 12(7): 1775-84.
[http://dx.doi.org/10.1016/j.nano.2016.04.008] [PMID: 27125435]

[36]   Jahangirian H, Lemraski EG, Webster TJ, Rafiee-Moghaddam R, Abdollahi Y. A review of drug delivery systems based on nanotechnology and green chemistry: green nanomedicine. Int J Nanomedicine 2017; 12: 2957-78.
[http://dx.doi.org/10.2147/IJN.S127683] [PMID: 28442906]

[37]   Francis G, Thombree R, Parekh F, Leksminarayan P. Bioinspired synthesis of gold nanoparticles Using Ficus benghalensis (Indian Banyan) Leaf Extract. Chemical. Sci Trans 2014; 3: 470-4.

[38]   Sri Ramkumar SR, Sivakumar N, Selvakumar G, *et al.* Green synthesized silver nanoparticles from Garcinia imberti bourd and their impact on root canal pathogens and HepG2 cell lines. RSC Advances

2017; 7: 34548-55.
[http://dx.doi.org/10.1039/C6RA28328D]

[39]   Constable DJC, Jimenez-Gonzales C, Henderson RK. Prospective on solvent use in the pharmaceutical industry. Org Process Res Dev 2007; 11: 133-7.
[http://dx.doi.org/10.1021/op060170h]

[40]   Zhang J, Cue BW. Green process chemistry in the pharmaceutical industry. Green Chem Lett Rev 2009; 2: 193-211.
[http://dx.doi.org/10.1080/17518250903258150]

[41]   Thomas V, Valavanidis A. 2013.

[42]   Patel M, Patel H, Patel O, Mewada S. Chemistry goes green: A review on current and future perspectives of pharmaceutical green chemistry. World J Pharm Med Res 2020; 6: 125-31.

[43]   Yogesh SS, Ravikumar VT, Anthony NS, Douglas LC. Applications of green chemistry in the manufacture of oligonucleotide drugs. Pure Appl Chem 2001; 73: 175-80.
[http://dx.doi.org/10.1351/pac200173010175]

[44]   Beaucage SL, Caruthers MH. In current protocols in nucleic acid chemistry. New York: Wiley 2000.

[45]   Krotz AH, Carty RL, Scozzari AN, *et al.* Large-Scale Synthesis of Antisense Oligonucleotides without Chlorinated Solvents. Org Process Res Dev 2000; 4: 190-3.
[http://dx.doi.org/10.1021/op990183d]

[46]   Anastas PT, Williamson TC. Green Chemistry Frontiers in Benign Chemical Syntheses and Process. New York: Oxford University Press 1998; p. 209.

[47]   Joncour R, Duguet N, Metay E, Ferriera A, Lemaire M. Amidation of phenol derivatives: A direct synthesis of parcetamol (acetaminophen) from hydroquinone. Electronic supplementary material for Green Chemistry. Green Chem 2014; 16: 2997-3002.
[http://dx.doi.org/10.1039/c4gc00166d]

[48]   Sheldon RA. Atom utilisation, *E* factors and the catalytic solution. C. R. Acad. Sci. Paris IIc. Chim 2000; 3: 541-51.

[49]   Mustapha S. Application of green chemistry in pharmaceutical industry. Int J Pharm Sci Res 2018; 3: 24-8.

[50]   Jennings S. A Green Process for the Synthesis of Quinapril Hydrochloride. Summary for the Presidential Green Chemistry Challenge Awards Program 2005. [Pfizer]

[51]   Jiménez-González C, Ponder CS, Broxterman QB, *et al.* Using the right green yardstick: why process mass intensity is used in the pharmaceutical industry to drive more sustainable processes. Org Process Res Dev 15: 912-7.
[http://dx.doi.org/10.1021/op200097d]

[52]   Badman DG, Jaffe ER. Blood and State Air Pollution; State of Knowledge and Research Needs Otolaryngology. Head Neck Surg 1996; 114: 205-8.
[http://dx.doi.org/10.1016/S0194-5998(96)70166-3] [PMID: 8637733]

[53]   Sadicova S, Buglanov AA, Tadzhieva ZA, *et al.* Indicators of iron metabolism and cellular immunity in healthy children and in those with iron deficiency anemia in relation to ecological condition 1990.

[54]   López CM, Piñeiro AE, Núñez N, Avagnina AM, Villaamil EC, Roses OE. Thyroid hormone changes in males exposed to lead in the Buenos Aires area (Argentina). Pharmacol Res 2000; 42(6): 599-602.
[http://dx.doi.org/10.1006/phrs.2000.0734] [PMID: 11058414]

[55]   Green Chemical Industry to Soar to USD 98.5 Billion by 2020, Navigant Research, Jun20, 2011

[56]   Prat D, Hayler J, Wells A. A survey of solvent selection guides. Green Chem 2014; 16: 4546-51.
[http://dx.doi.org/10.1039/C4GC01149J]

[57]   Henderson RK, Jimenez-Gonzalez C, Constable DJC, *et al.* Expanding GSK's solvent selection guide

–Embedding sustainability into solvent selection starting at medicinal chemistry. Green Chem 2011; 13: 854-62.
[http://dx.doi.org/10.1039/c0gc00918k]

[58]   Kua YL, Gan S, Morris A, *et al.* Ethyll actate as a potential green solvent to extract hydrophilic (polar) and lipophilic (non-polar) phyto nutrients simultaneously from fruit and vegetable by-products. Sustainable Chem Pharm 2016; 4: 21-31.
[http://dx.doi.org/10.1016/j.scp.2016.07.003]

[59]   Li C-J, Chan T-K. Organic reactions in aqueous media. New York, NY: Wiley 1997.

[60]   Clifford AA. Fundamentals of supercritical fluids. Oxford, UK: Oxford University Press 1998.

[61]   Jessop PG, Subramaniam B. Gas-expanded liquids. Chem Rev 2007; 107(6): 2666-94.
[http://dx.doi.org/10.1021/cr040199o] [PMID: 17564482]

[62]   Wasserscheid P, Welton T, Eds. Ionic liquids in synthesis. 2nd ed., Weinheim, Germany: Wiley-VCH 2008.

[63]   Chandrasekhar S, Narsihmulu Ch, Sultana SS, Reddy NR. Poly(ethylene glycol) (PEG) as a reusable solvent medium for organic synthesis. Application in the Heck reaction. Org Lett 2002; 4(25): 4399-401.
[http://dx.doi.org/10.1021/ol0266976] [PMID: 12465897]

[64]   Mathers RT, McMahon KC, Damodaran K, *et al.* Ring-opening metathesis polymerizations in D-limonene: a renewable polymerization solvent and chain transfer agent for the synthesis of alkene macromonomers. Macromolecules 2006; 39: 8982-6.
[http://dx.doi.org/10.1021/ma061699h]

[65]   Curzons AD, Constable DC, Cunningham VL. 1999 Solvent selection guide: a guide to the integration of environmental, health and safety criteria into the selection of solvents. Clean Prod Process 1999; 1: 82-90.

[66]   Jiménez-González C, Curzons AD, Constable DC, *et al.* 2005 Expanding GSK's Solvent Selection Guide—application of life cycle assessment to enhance solvent selections. Clean Technol Environ Policy 2005; 7: 42-50.
[http://dx.doi.org/10.1007/s10098-004-0245-z]

[67]   Alfonsi K, Colberg J, Dunn PJ, *et al.* Green chemistry tools to influence a medicinal chemistry and research chemistry based organisation. Green Chem 2008; 10: 31-6.
[http://dx.doi.org/10.1039/B711717E]

[68]   Prat D, Pardigon O, Flemming HW, *et al.* Sanofi's solvent selection guide: a step toward more sustainable processes. Org Process Res Dev 2013; 17: 1517-25.
[http://dx.doi.org/10.1021/op4002565]

[69]   Smita T, Falfuni M. Green chemistry: A tool in pharmaceutical chemistry. NHL J Med Sci 2012; 1: 7-13.

[70]   Morsch LA, Deak L, Tiburzi D, *et al.* Green aqueous Wittig reaction: Teaching green chemistry in organic teaching laboratories. J Chem Educ 2014; 91: 611-4.
[http://dx.doi.org/10.1021/ed400408k]

[71]   Sobral AJFN. Synthesis of meso-diethyl-2,2′-dipyrromethane in water. J Chem Educ 2006; 83: 1665-6.
[http://dx.doi.org/10.1021/ed083p1665]

[72]   Li Q, Yan Y, Wang X, *et al.* Water as a green solvent for efficient synthesis of isocoumarins through microwave-accelerated and Rh/Cu-catalyzed C–H/O–H bond functionalization. RSC Advances 2013; 3: 23402-8.
[http://dx.doi.org/10.1039/c3ra43175d]

[73]   Pirrung MC, Sarma KD. Multicomponent reactions are accelerated in water. J Am Chem Soc 2004;

126(2): 444-5.
[http://dx.doi.org/10.1021/ja038583a] [PMID: 14719923]

[74]   Yorimitsu H, Nakamura T, Shinokubo H, *et al.* Powerful solvent effect of water in radical reaction: Triethylborane-induced atom-transfer radical cyclization in water. J Am Chem Soc 2000; 122: 11041-7.
[http://dx.doi.org/10.1021/ja0014281]

[75]   Pinto AC, Lapis AAM, Silva BV, *et al.* Pronounced ionic liquid effect in the synthesis of biologically active isatin-3-oxime derivatives under acid catalysis. Tetrahedron Lett 2008; 49: 5639-41.
[http://dx.doi.org/10.1016/j.tetlet.2008.07.067]

[76]   Jose IG, Garcia-Martin H, Mayoral JA, *et al.* Green solvents from Glycerol, synthesis and physico-chemical properties of alkyl glycerol ethers. Green Chem 2010; 12: 426-34.
[http://dx.doi.org/10.1039/b923631g]

[77]   Safaei HR, Shekouhy M, Rahmanpur S, *et al.* Glycerol as a biodegradable and reusable promoting medium for the catalyst-free one-pot three-component synthesis of 4H-pyrans. Green Chem 2012; 14: 1696-704.
[http://dx.doi.org/10.1039/c2gc35135h]

[78]   Gu Y, Barrault J, Jerome F. Glycerol as an efficient promoting medium for organic reactions. Adv Synth Catal 2008; 350: 2007-12.
[http://dx.doi.org/10.1002/adsc.200800328]

[79]   Garcia JI, Garcia-Marin H, Mayoral JA, *et al.* Green solvents from glycerol. Synthesis and physico-chemical properties of alkyl glycerol ethers. Green Chem 2010; 12: 426-34.
[http://dx.doi.org/10.1039/b923631g]

[80]   Draye M, Chatel G, Duwald R. Ultrasound for Drug Synthesis: A Green Approach. Pharmaceuticals (Basel) 2020; 13(2): 23.
[http://dx.doi.org/10.3390/ph13020023] [PMID: 32024033]

[81]   Oak RS, Clifford AA, Rayner CA. The use of supercritical fluids in synthetic organic. Chem Soc Perkin Trans 2001; 1: 917-41.

[82]   Menges N, Şahin E. Metal- and base-free combinatorial reaction for C-acylation of 1,3-diketo compounds in vegetable oil: The effect of natural oil. ACS Sustain Chem& Eng 2014; 2: 226-30.
[http://dx.doi.org/10.1021/sc400281h]

[83]   Yang Z, Pan W. Ionic liquids: Green solvents for nonaqueous biocatalysis. Enzyme Microb Technol 2005; 37: 19-28.
[http://dx.doi.org/10.1016/j.enzmictec.2005.02.014]

[84]   Lopes EF, Gonçalves LC, Vinueza JCG, *et al.* DES as a green solvent to prepare 1,2-bis-organylseleno alkenes. Scope and limitations. Tetrahedron Lett 2015; 56: 6890-5.
[http://dx.doi.org/10.1016/j.tetlet.2015.10.095]

SUBJECT INDEX

A

Absorption 17, 31, 36, 59, 60, 168, 171, 186, 202, 211, 213, 216, 217
  distribution, metabolism, and excretion (ADME) 17, 36, 168, 202, 211, 217
  oral 211
  rate 60, 216
Acid(s) 105, 122, 150, 151, 160, 167, 171, 174, 175, 190, 205, 217, 223, 231, 232, 235, 237
  acetic 232
  acetoacetic 235
  aristolochic 150, 151
  benzene sulphonic 175
  benzoic 174
  Gallic 171
  hydrochloric 105
  lactic 223
  morphine sulphuric 175
  nitric 231
  nitrophenanthrene carboxylic 150
  nucleic 122, 167, 217
  oxalic 237
  protein-nucleic 205
  salicylic 174, 235
  triterpenoid 160
Acquired immunodeficiency syndrome 25
Action 168, 184, 229
  enzymatic 229
  pharmacological 168, 184
Activates estrogen receptor 64
Activation 5, 30, 51, 150, 152, 173, 186, 188
  blocks lymphocytes 30
  drug-mediated receptor 186
  steam 5
  transcriptional 51
Activator protein 52
Active 4, 26, 31, 35, 179, 192
  chemical compounds 26, 31
  pain killer 179
  pharmaceutical ingredient (APIs) 4, 35, 192

Activities 43, 48, 81, 166, 180, 187, 216
  anthropogenic 81
  antiestrogenic 48
  anti-inflammatory 180
  hormonal 43
  inflammatory 166
  intrinsic 187
  metabolic 216
Acute regulatory proteins 44
Adenocarcinoma 148
Adipose tissues 77, 80, 81, 84, 93
Advanced oxidation process (AOPs) 92, 93
Agency for toxic substances and disease registry (ATSDR) 2, 8, 102
Air 4, 61, 62, 76, 79, 80, 84, 89, 102, 103, 153, 154, 233
  atmospheric 76
  contaminated 89
  pollution 233
Algal degradation 92
Analysis 216, 228
  metabolic pathway 216
  retrosynthetic 228
Androgen receptor (AR) 57, 126
Angiotensin-converting enzyme (ACE) 32
Anticancer properties 160
Antidiabetic agent 22
Antiestrogenic compounds 48
Anti-inflammatory 22, 28
  action 28
  agent 22
Aplastic anaemia 233
Apoptosis 122, 147, 148, 158, 160
  signals 148
Ariens-stephensen theory 187
Aromatase enzyme 44
Aspergillus 93, 149, 150
  awamori 93
  flavus 149, 150
Asthma 26, 28, 30
Atherosclerosis 24

Tahmeena Khan, Abdul Rahman Khan, Saman Raza, Iqbal Azad and Alfred J. Lawrence (Eds.)